Loren Eiseley

THE

IMMENSE

JOURNEY

VINTAGE BOOKS

A DIVISION OF RANDOM HOUSE

New York

THE IMMENSE JOURNEY

dedicated to the memory of
CLYDE EDWIN EISELEY,
who lies in the grass of the prairie frontier
but is not forgotten by his son

The author wishes to thank the editors of the *American Scholar*, *Harper's Magazine*, and the *Scientific American* for permission to reprint material which appeared separately in those publications. He would like, also, to express his gratitude to the Wenner-Gren Foundation for Anthropological Research for providing the leisure from professional duties during which a major number of the essays in this book were written.

CONTENTS

THE IMMENSE JOURNEY

"Man can not afford to be a naturalist, to look at Nature directly, but only with the side of his eye. He must look through and beyond her."

HENRY DAVID THOREAU

"Unless all existence is a medium of revelation, no particular revelation is possible. . . ."

WILLIAM TEMPLE

THE SLIT

🌲

Some lands are flat and grass-covered, and smile so evenly up at the sun that they seem forever youthful, untouched by man or time. Some are torn, ravaged and convulsed like the features of profane old age. Rocks are wrenched up and exposed to view; black pits receive the sun but give back no light.

It was to such a land I rode, but I rode to it across a sunlit, timeless prairie over which nothing passed but antelope or a wandering bird. On the verge where that prairie halted before a great wall of naked sandstone and clay, I came upon the Slit. A narrow crack worn by some descending torrent had begun secretly, far back in the prairie grass, and worked itself deeper and deeper into the fine sandstone that led by devious channels into the broken waste beyond. I rode back along the crack to a spot where I could descend into it, dismounted, and left my horse to graze.

The crack was only about body-width and, as I worked my way downward, the light turned dark and green from the overhanging grass. Above me the sky became a narrow slit of distant blue, and the sandstone was cool to my hands on either side. The Slit was a little sinister—like an open grave, assuming the dead were enabled to take one last look—for over me the sky seemed already as far off as some future century I would never see.

I ignored the sky, then, and began to concentrate on the sandstone walls that had led me into this place. It was tight and tricky work, but that cut was a perfect cross section through perhaps ten million years of time. I hoped to find at least a bone, but I was not quite prepared for the sight I finally came upon. Staring straight out at me, as I slid farther and deeper into the green twilight, was a skull embedded in the solid sandstone. I had come at just the proper moment when it was fully to be seen, the white bone gleaming there in a kind of ashen splendor, water worn, and about to be ground away in the next long torrent.

It was not, of course, human. I was deep, deep below the time of man in a remote age near the beginning of the reign of mammals. I squatted on my heels in the narrow ravine, and we stared a little blankly at each other, the skull and I. There were marks of generalized primitiveness in that low, pinched brain case and grinning jaw that marked it as lying far back along those converging roads where, as I shall have occasion to establish elsewhere, cat and man and weasel must leap into a single shape.

It was the face of a creature who had spent his days following his nose, who was led by instinct rather than memory, and whose power of choice was very small. Though he was not a man, nor a direct human ancestor, there was yet about him, even in the bone, some trace of that low, snuffling world out of which our forebears had so recently emerged. The skull lay tilted in such a manner that it stared, sightless, up at me as though I, too, were already caught a few feet above him in the strata and, in my turn, were staring upward at that strip of sky which the ages were carrying farther away from me beneath the tumbling debris of falling mountains. The creature had never lived to see a man, and I, what was it I was never going to see?

I restrained a panicky impulse to hurry upward after that receding sky that was outlined above the Slit. Probably, I thought, as I patiently began the task of chiseling into the stone around the skull, I would never again excavate a fossil under conditions which led to so vivid an impression that I was already one myself. The truth is that we are all potential fossils still carrying within our bodies the crudities of former existences, the marks of a world in which living creatures flow with little more consistency than clouds from age to age.

As I tapped and chiseled there in the foundations of the world, I had ample time to consider the cunning manipulability of the human fingers. Experimentally I crooked one of the long slender bones. It might have been silica, I thought, or aluminum, or iron—the cells would have made it possible. But no, it is calcium, carbonate of lime. Why? Only because of its history. Elements more numerous than calcium in the earth's crust could have been used to build the skeleton. Our history is the reason—we came from the water. It was there the cells took the lime habit, and they kept it after we came ashore.

It is not a bad symbol of that long wandering, I thought again—the human hand that has been fin and scaly reptile foot and furry paw. If a stone should fall (I cocked an eye at the leaning shelf above my head and waited, fatalistically) let the bones lie here with their message, for those who might decipher it, if they come down late among us from the stars.

Above me the great crack seemed to lengthen.

Perhaps there is no meaning in it at all, the thought went on inside me, save that of journey itself, so far as men can see. It has altered with the chances of life, and the chances brought us here; but it was a good journey

—long, perhaps—but a good journey under a pleasant sun. Do not look for the purpose. Think of the way we came and be a little proud. Think of this hand—the utter pain of its first venture on the pebbly shore.

Or consider its later wanderings.

I ceased my tappings around the sand-filled sockets of the skull and wedged myself into a crevice for a smoke. As I tamped a load of tobacco into my pipe, I thought of a town across the valley that I used sometimes to visit, a town whose little inhabitants never welcomed me. No sign points to it and I rarely go there any more. Few people know about it and fewer still know that in a sense we, or rather some of the creatures to whom we are related, were driven out of it once, long ago. I used to park my car on a hill and sit silently observant, listening to the talk ringing out from neighbor to neighbor, seeing the inhabitants drowsing in their doorways, taking it all in with nostalgia—the sage smell on the wind, the sunlight without time, the village without destiny. We can look, but we can never go back. It is prairie-dog town.

"Whirl is king," said Aristophanes, and never since life began was Whirl more truly king than eighty million years ago in the dawn of the Age of Mammals. It would come as a shock to those who believe firmly that the scroll of the future is fixed and the roads determined in advance, to observe the teetering balance of earth's history through the age of the Paleocene. The passing of the reptiles had left a hundred uninhabited life zones and a scrambling variety of newly radiating forms. Unheard-of species of giant ground birds threatened for a moment to dominate the earthly scene. Two separate orders of life contended at slightly different intervals for the pleasant grasslands—for the seeds and the sleepy burrows in the sun.

Sometimes, sitting there in the mountain sunshine above prairie-dog town, I could imagine the attraction of that open world after the fern forest damp or the croaking gloom of carboniferous swamps. There by a tree root I could almost make him out, that shabby little Paleocene rat, eternal tramp and world wanderer, father of all mankind. He ruffed his coat in the sun and hopped forward for a seed. It was to be a long time before he would be seen on the grass again, but he was trying to make up his mind. For good or ill there was to be one more chance, but that chance was fifty million years away.

Here in the Paleocene occurred the first great radiation of the placental mammals, and among them were the earliest primates—the zoological order to which man himself belongs. Today, with a few unimportant exceptions, the primates are all arboreal in habit except man. For this reason we have tended to visualize all of our remote relatives as tree dwellers. Recent discoveries, however, have begun to alter this one-sided picture. Before the rise of the true rodents, the highly successful order to which present-day prairie dogs and chipmunks belong, the environment which they occupy had remained peculiarly open to exploitation. Into this zone crowded a varied assemblage of our early relatives.

"In habitat," comments one scholar, "many of these early primates may be thought of as the rats of the Paleocene. With the later appearance of true rodents, the primate habitat was markedly restricted." The bone hunters, in other words, have succeeded in demonstrating that numerous primates reveal a remarkable development of rodent-like characteristics in the teeth and skull during this early period of mammalian evolution. The movement is progressive and distributed in several different groups. One form, although that of a true primate, shows similarities to the modern kangaroo rat, which is, of course, a rodent. There is little doubt that it was a burrower.

It is this evidence of a lost chapter in the history of our kind that I used to remember on the sunny slope above prairie-dog town, and that enables me to say in a somewhat figurative fashion that we were driven out of it once ages ago. We are not, except very remotely as mammals, related to prairie dogs. Nevertheless, through several million years of Paleocene time, the primate order, instead of being confined to trees, was experimenting to some extent with the same grassland burrowing life that the rodents later perfected. The success of these burrowers crowded the primates out of this environment and forced them back into the domain of the branches. As a result, many primates, by that time highly specialized for a ground life, became extinct.

In the restricted world of the trees, a "refuge area," as the zoologist would say, the others lingered on in diminished numbers. Our ancient relatives, it appeared, were beaten in their attempt to expand upon the ground; they were dying out in the temperate zone, and their significance as a widespread and diversified group was fading. The shabby pseudo-rat I had seen ruffling his coat to dry after the night damps of the reptile age, had ascended again into the green twilight of the rain forest. The chatterers with the ever-growing teeth were his masters. The sunlight and the grass belonged to them.

It is conceivable that except for the invasion of the rodents, the primate line might even have abandoned the trees. We might be there on the grass, you and I, barking in the high-plains sunlight. It is true we came back in fifty million years with the cunning hands and the eyes that the tree world gave us, but was it victory? Once more in memory I saw the high blue evening fall sleepily upon that village, and once more swung the car to leave, lifting, as I always did, a figurative lantern to some ambiguous crossroads sign within my brain. The pointing arms were nameless and nameless were the distances to which they pointed. One took one's choice.

I ceased my daydreaming then, squeezed myself out of the crevice, shook out my pipe, and started chipping once more, the taps sounding along the inward-leaning walls of the Slit like the echo of many footsteps ascending and descending. I had come a long way down since morning; I had projected myself across a dimension I was not fitted to traverse in the flesh. In the end I collected my tools and climbed painfully up through the colossal debris of ages. When I put my hands on the surface of the crack I looked all about carefully in a sudden anxiety that it might not be a grazing horse that I would see.

He had not visibly changed, however, and I mounted in some slight trepidation and rode off, having a memory for a camp—if I had gotten a foot in the right era —which should lie somewhere over to the west. I did not, however, escape totally from that brief imprisonment.

Perhaps the Slit, with its exposed bones and its far-off vanishing sky, has come to stand symbolically in my mind for a dimension denied to man, the dimension of time. Like the wistaria on the garden wall he is rooted in his particular century. Out of it—forward or backward—he cannot run. As he stands on his circumscribed pinpoint of time, his sight for the past is growing longer, and even the shadowy outlines of the galactic future are growing clearer, though his own fate he cannot yet see. Along the dimension of time, man, like the rooted vine in space, may never pass in person. Considering the innumerable devices by which the mindless root has evaded the limitations of its own stability, however, it may well be that man himself is slowly achieving powers over a new dimension—a dimension capable of presenting him with a wisdom he has barely begun to discern.

Through how many dimensions and how many media will life have to pass? Down how many roads among the stars must man propel himself in search of the final secret? The journey is difficult, immense, at times impossible, yet that will not deter some of us from attempting it. We cannot know all that has happened in the past, or the reason for all of these events, any more than we can with surety discern what lies ahead. We have joined the caravan, you might say, at a certain point; we will travel as far as we can, but we cannot in one lifetime see all that we would like to see or learn all that we hunger to know.

The reader who would pursue such a journey with me is warned that the essays in this book have not been brought together as a guide but are offered rather as a somewhat unconventional record of the prowlings of one mind which has sought to explore, to understand, and to enjoy the miracles of this world, both in and out of science. It is, without doubt, an inconsistent record in many ways, compounded of fear and hope, for it has grown out of the seasonal jottings of a man preoccupied with time. It involves, I see now as I come to put it together, the four ancient elements of the Greeks: mud and the fire within it we call life, vast waters, and something—space, air, the intangible substance of hope which at the last proves unanalyzable by science, yet out of which the human dream is made.

Forward and backward I have gone, and for me it has been an immense journey. Those who accompany me need not look for science in the usual sense, though I have done all in my power to avoid errors in fact. I have given the record of what one man thought as he pursued research and pressed his hands against the confining walls of scientific method in his time. It is not, I must confess at the outset, an account of discovery so much as a confession of ignorance and of the final illumination that sometimes comes to a man when he is no longer careful of his pride. In the last three chapters of the book I have tried to put down such miracles as can be evoked from common earth. But men see differently. I can at best report only from my own wilderness. The important thing is that each man possess such a wilderness and that he consider what marvels are to be observed there.

Finally, I do not pretend to have set down, in Baconian terms, a true, or even a consistent model of the universe. I can only say that here is a bit of my personal universe, the universe traversed in a long and uncompleted journey. If my record, like those of the sixteenth-century voyagers, is confused by strange beasts or monstrous thoughts or sights of abortive men, these are no more than my eyes saw or my mind conceived. On the world island we are all castaways, so that what is seen by one may often be dark or obscure to another.

THE FLOW
OF THE RIVER

✿

If there is magic on this planet, it is contained in water. Its least stir even, as now in a rain pond on a flat roof opposite my office, is enough to bring me searching to the window. A wind ripple may be translating itself into life. I have a constant feeling that some time I may witness that momentous miracle on a city roof, see life veritably and suddenly boiling out of a heap of rusted pipes and old television aerials. I marvel at how suddenly a water beetle has come and is submarining there in a spatter of green algae. Thin vapors, rust, wet tar and sun are an alembic remarkably like the mind; they throw off odorous shadows that threaten to take real shape when no one is looking.

Once in a lifetime, perhaps, one escapes the actual confines of the flesh. Once in a lifetime, if one is lucky, one so merges with sunlight and air and running water that whole eons, the eons that mountains and deserts know, might pass in a single afternoon without discomfort. The mind has sunk away into its beginnings among old roots and the obscure tricklings and movings that stir inanimate things. Like the charmed fairy circle into which a man once stepped, and upon emergence learned that a whole century had passed in a single night, one can never quite define this secret; but it has something to do, I am sure, with common water. Its substance reaches everywhere; it touches the past and prepares the future; it moves under the poles and wanders thinly in the heights of air. It can assume forms of exquisite perfection in a snowflake, or strip the living to a single shining bone cast up by the sea.

Many years ago, in the course of some scientific investigations in a remote western county, I experienced, by chance, precisely the sort of curious absorption by water—the extension of shape by osmosis—at which I have been hinting. You have probably never experienced in yourself the meandering roots of a whole watershed or felt your outstretched fingers touching, by some kind of clairvoyant extension, the brooks of snow-line glaciers at the same time that you were flowing toward the Gulf over the eroded debris of worn-down mountains. A poet, MacKnight Black, has spoken of being "limbed . . . with waters gripping pole and pole." He had the idea, all right, and it is obvious that these sensations are not unique, but they are hard to come by; and the sort of extension of the senses that people will accept when they put their ear against a sea shell, they will smile at in the confessions of a bookish professor. What makes it worse is the fact that because of a traumatic experience in childhood, I am not a swimmer, and am inclined to be timid before any large body of water. Perhaps it was just this, in a way, that contributed to my experience.

As it leaves the Rockies and moves downward over the high plains towards the Missouri, the Platte River is a curious stream. In the spring floods, on occasion, it can be a mile-wide roaring torrent of destruction, gulping farms and bridges. Normally, however, it is a rambling, dispersed series of streamlets flowing erratically over great sand and gravel fans that are, in part, the remnants of a mightier Ice Age stream bed. Quicksands and shifting islands haunt its waters. Over it the prairie suns beat mercilessly throughout the summer. The Platte, "a mile wide and an inch deep," is a refuge for any heat-weary pilgrim along its shores. This is particularly true on the high plains before its long march by the cities begins.

The reason that I came upon it when I did, breaking through a willow thicket and stumbling out through ankle-deep water to a dune in the shade, is of no concern to this narrative. On various purposes of science I have ranged over a good bit of that country on foot, and I know the kinds of bones that come gurgling up through the gravel pumps, and the arrowheads of shining chalcedony that occasionally spill out of water-loosened sand. On that day, however, the sight of sky and willows and the weaving net of water murmuring a little in the shallows on its way to the Gulf stirred me, parched as I was with miles of walking, with a new idea: I was going to float. I was going to undergo a tremendous adventure.

The notion came to me, I suppose, by degrees. I had shed my clothes and was floundering pleasantly in a hole among some reeds when a great desire to stretch out and go with this gently insistent water began to pluck at me. Now to this bronzed, bold, modern generation, the struggle I waged with timidity while standing there in knee-deep water can only seem farcical; yet actually for me it was not so. A near-drowning accident in childhood had scarred my reactions; in addition to the fact that I was a nonswimmer, this "inch-deep river" was treacherous with holes and quicksands. Death was not precisely infrequent along its wandering and illusory channels. Like all broad wastes of this kind, where neither water nor land quite prevails, its thickets were lonely and untraversed. A man in trouble would cry out in vain.

I thought of all this, standing quietly in the water,

feeling the sand shifting away under my toes. Then I lay back in the floating position that left my face to the sky, and shoved off. The sky wheeled over me. For an instant, as I bobbed into the main channel, I had the sensation of sliding down the vast tilted face of the continent. It was then that I felt the cold needles of the alpine springs at my fingertips, and the warmth of the Gulf pulling me southward. Moving with me, leaving its taste upon my mouth and spouting under me in dancing springs of sand, was the immense body of the continent itself, flowing like the river was flowing, grain by grain, mountain by mountain, down to the sea. I was streaming over ancient sea beds thrust aloft where giant reptiles had once sported; I was wearing down the face of time and trundling cloud-wreathed ranges into oblivion. I touched my margins with the delicacy of a crayfish's antennae, and felt great fishes glide about their work.

I drifted by stranded timber cut by beaver in mountain fastnesses; I slid over shallows that had buried the broken axles of prairie schooners and the mired bones of mammoth. I was streaming alive through the hot and working ferment of the sun, or oozing secretively through shady thickets. I *was* water and the unspeakable alchemies that gestate and take shape in water, the slimy jellies that under the enormous magnification of the sun writhe and whip upward as great barbeled fish mouths, or sink indistinctly back into the murk out of which they arose. Turtle and fish and the pinpoint chirpings of individual frogs are all watery projections, concentrations—as man himself is a concentration—of that indescribable and liquid brew which is compounded in varying proportions of salt and sun and time. It has appearances, but at its heart lies water, and as I was finally edged gently against a sand bar and dropped like any log, I tottered as I rose. I knew once more the body's revolt against emergence into the harsh and un-supporting air, its reluctance to break contact with that mother element which still, at this late point in time, shelters and brings into being nine tenths of everything alive.

As for men, those myriad little detached ponds with their own swarming corpuscular life, what were they but a way that water has of going about beyond the reach of rivers? I, too, was a microcosm of pouring rivulets and floating driftwood gnawed by the mys-terious animalcules of my own creation. I was three fourths water, rising and subsiding according to the hollow knocking in my veins: a minute pulse like the eternal pulse that lifts Himalayas and which, in the fol-lowing systole, will carry them away.

Thoreau, peering at the emerald pickerel in Walden Pond, called them "animalized water" in one of his moments of strange insight. If he had been possessed of the geological knowledge so laboriously accumulated since his time, he might have gone further and amusedly detected in the planetary rumblings and eructations which so delighted him in the gross habits of certain frogs, signs of that dark interior stress which has reared sea bottoms up to mountainous heights. He might have developed an acute inner ear for the sound of the surf on Cretaceous beaches where now the wheat of Kansas rolls. In any case, he would have seen, as the long trail of life was unfolded by the fossil hunters, that his animalized water had changed its shapes eon by eon to the beating of the earth's dark millennial heart. In the swamps of the low continents, the amphibians had flourished and had their day; and as the long skyward swing—the isostatic response of the crust—had come about, the era of the cooling grasslands and mammalian life had come into being.

A few winters ago, clothed heavily against the weather, I wandered several miles along one of the tributaries of that same Platte I had floated down years before. The land was stark and ice-locked. The rivulets were frozen, and over the marshlands the willow thickets made such an array of vertical lines against the snow that tramping through them produced strange optical illusions and dizziness. On the edge of a frozen backwater, I stopped and rubbed my eyes. At my feet a raw prairie wind had swept the ice clean of snow. A peculiar green object caught my eye; there was no mistaking it.

Staring up at me with all his barbels spread pathetically, frozen solidly in the wind-ruffled ice, was a huge familiar face. It was one of those catfish of the twisting channels, those dwellers in the yellow murk, who had been about me and beneath me on the day of my great voyage. Whatever sunny dream had kept him paddling there while the mercury plummeted downward and that Cheshire smile froze slowly, it would be hard to say. Or perhaps he was trapped in a blocked channel and had simply kept swimming until the ice contracted around him. At any rate, there he would lie till the spring thaw.

At that moment I started to turn away, but something in the bleak, whiskered face reproached me, or perhaps it was the river calling to her children. I termed it science, however—a convenient rational phrase I reserve for such occasions—and decided that I would cut the fish out of the ice and take him home. I had no intention of eating him. I was merely struck by a sudden impulse to test the survival qualities of high-plains fishes, particularly fishes of this type who get themselves immured in oxygenless ponds or in cut-off oxbows buried in winter drifts. I blocked him out as gently as possible and dropped him, ice and all, into a collecting can in the car. Then we set out for home.

Unfortunately, the first stages of what was to prove a remarkable resurrection escaped me. Cold and tired after a long drive, I deposited the can with its melting water and ice in the basement. The accompanying corpse I anticipated I would either dispose of or dissect on the following day. A hurried glance had revealed no signs of life.

To my astonishment, however, upon descending into the basement several hours later, I heard stirrings in the receptacle and peered in. The ice had melted. A vast pouting mouth ringed with sensitive feelers confronted me, and the creature's gills labored slowly. A thin stream of silver bubbles rose to the surface and popped. A fishy eye gazed up at me protestingly.

"A tank," it said. This was no Walden pickerel. This was a yellow-green, mud-grubbing, evil-tempered inhabitant of floods and droughts and cyclones. It was the selective product of the high continent and the waters that pour across it. It had outlasted prairie blizzards that left cattle standing frozen upright in the drifts.

"I'll get the tank," I said respectfully.

He lived with me all that winter, and his departure was totally in keeping with his sturdy, independent character. In the spring a migratory impulse or perhaps sheer boredom struck him. Maybe, in some little lost corner of his brain, he felt, far off, the pouring of the mountain waters through the sandy coverts of the Platte. Anyhow, something called to him, and he went. One night when no one was about, he simply jumped out of his tank. I found him dead on the floor next morning. He had made his gamble like a man—or, I should say, a fish. In the proper place it would not have been a fool's gamble. Fishes in the drying shallows of intermittent prairie streams who feel their confinement and have the impulse to leap while there is yet time may regain the main channel and survive. A million ancestral years had gone into that jump, I thought as I looked at him, a million years of climbing through prairie sunflowers and twining in and out through the pillared legs of drinking mammoth.

"Some of your close relatives have been experimenting with air breathing," I remarked, apropos of nothing, as I gathered him up. "Suppose we meet again up there in the cottonwoods in a million years or so."

I missed him a little as I said it. He had for me the kind of lost archaic glory that comes from the water brotherhood. We were both projections out of that timeless ferment and locked as well in some greater unity that lay incalculably beyond us. In many a fin and reptile foot I have seen myself passing by—some part of myself, that is, some part that lies unrealized in the momentary shape I inhabit. People have occasionally written me harsh letters and castigated me for a lack of faith in man when I have ventured to speak of this matter in print. They distrust, it would seem, all shapes and thoughts but their own. They would bring God into the compass of a shopkeeper's understanding and confine Him to those limits, lest He proceed to some unimaginable and shocking act—create perhaps, as a casual afterthought, a being more beautiful than man. As for me, I believe nature capable of this, and having been part of the flow of the river, I feel no envy —any more than the frog envies the reptile or an ancestral ape should envy man.

Every spring in the wet meadows and ditches I hear a little shrilling chorus which sounds for all the world like an endlessly reiterated "We're here, we're here, we're here." And so they are, as frogs, of course. Confident little fellows. I suspect that to some greater ear than ours, man's optimistic pronouncements about his role and destiny may make a similar little ringing sound that travels a small way out into the night. It is only its nearness that is offensive. From the heights of a mountain, or a marsh at evening, it blends, not too badly, with all the other sleepy voices that, in croaks or chirrups, are saying the same thing.

After a while the skilled listener can distinguish man's noise from the katydid's rhythmic assertion, allow for the offbeat of a rabbit's thumping, pick up the autumnal monotone of crickets, and find in all of them a grave pleasure without admitting any to a place of preëminence in his thoughts. It is when all these voices cease and the waters are still, when along the frozen river nothing cries, screams or howls, that the enormous mindlessness of space settles down upon the soul. Somewhere out in that waste of crushed ice and reflected stars, the black waters may be running, but they appear to be running without life toward a destiny in which the whole of space may be locked in some silvery winter of dispersed radiation.

It is then, when the wind comes straitly across the barren marshes and the snow rises and beats in endless waves against the traveler, that I remember best, by some trick of the imagination, my summer voyage on the river. I remember my green extensions, my catfish nuzzlings and minnow wrigglings, my gelatinous materializations out of the mother ooze. And as I walk on through the white smother, it is the magic of water that leaves me a final sign.

Men talk much of matter and energy, of the struggle for existence that molds the shape of life. These things exist, it is true; but more delicate, elusive, quicker than the fins in water, is that mysterious principle known as "organization," which leaves all other mysteries concerned with life stale and insignificant by comparison. For that without organization life does not persist is obvious. Yet this organization itself is not strictly the product of life, nor of selection. Like some dark and passing shadow within matter, it cups out the eyes' small windows or spaces the notes of a meadow lark's song in the interior of a mottled egg. That principle— I am beginning to suspect—was there before the living in the deeps of water.

The temperature has risen. The little stinging needles have given way to huge flakes floating in like white leaves blown from some great tree in open space. In the car, switching on the lights, I examine one intricate crystal on my sleeve before it melts. No utilitarian philosophy explains a snow crystal, no doctrine of use or disuse. Water has merely leapt out of vapor and thin nothingness in the night sky to array itself in form. There is no logical reason for the existence of a snowflake any more than there is for evolution. It is an apparition from that mysterious shadow world beyond nature, that final world which contains—if anything contains—the explanation of men and catfish and green leaves.

THE GREAT DEEPS

There is a night world that few men have entered and from whose greatest depths none have returned alive—the abyssal depths of the sea. Darwin's associates dreamed of its hovering and intangible shapes as possibly those of the lost Paleozoic world. The great naturalist himself pleaded with outbound voyagers: "Urge the use of the dredge in the tropics; how little or nothing we know of the limit of life downward in the hot seas."

Anything that has been supposedly dead for a hundred million years—anything that no living eye has beheld except in the chalks of vanished geological epochs—is monstrous when you find it alive and pulsing in your hand. But this was the experience of Sir Charles Thomson, one of the first explorers of the North Atlantic sea bed. Few men in the years since have laid hands or eyes upon living denizens of the fossil kingdom, and this was an adventure no man could forget. The discovery influenced indirectly the formation of the world's largest oceanographic expedition and made Sir Charles its leader. Speaking of his find, long afterwards, he said: "It was like a little round red cake. And like a little round red cake, it began to pant there in my hand. Curious undulations were passing through it and I had to summon up all my resolution before handling the weird little monster."

Now to the ordinary man that little round red cake would have been a sea urchin, and whether it panted would have meant nothing at all except that it was alive. Nevertheless, the man in the street would have been wrong. The fact *was* monstrous and the little red sea urchin more startling still. Even the "panting" had significance. No living sea urchin had ever been observed in such a performance. The known forms were all too rigid. The undulations of this little beast were a sure sign of its relationship to a more leathery and flexible ancestral group.

As a living fossil it had been dredged out of the North Atlantic sea bed almost a solid mile below the surface. A mile today is not a great depth compared with the six-mile depression of the Tuscarora Deep, but in the

sixties of the last century—Sir Charles Thomson's time
—it was below the level at which life was generally
supposed to exist. Anything below three hundred
fathoms was Azoic, lifeless—so wrote Edward Forbes,
the first great oceanographer of the eighteen forties.
Like many pioneers he was destined to be proved
wrong, yet looking back, it is possible to sympathize.
The cold, the dark, the pressure of those unknown
depths was frightening to contemplate. The human
mind shied subconsciously away from the notion that
even here sentient beings had groped their way down
into the primeval slime of the sea floor. It was the world
of the abyss, supposedly as lifeless as the earth's first
midnight.

Today we know that the abyss is haunted. Through
it drift luminous jack-o'-lantern faces with wolf-trap
mouths and meager bodies, as though a head floating in
that enormous darkness were more important than a
body, which could almost be dispensed with in the
lean economy of the night. It is a world of delicately
groping yard-long antennae, or of great staring eyes
that can pick up remote pinpoints of light and follow
them through the restless luminescence of a firefly
darkness. To Sir Charles Thomson, however, the abyss
was more than haunted. It was the world of the past.

The fascination of lost worlds has long preoccupied
humanity. It is inevitable that transitory man, student

of the galaxies and computer of light-years, should entertain nostalgic yearnings for some island outside of time, some Avalon untouched by human loss. Even the scholar has not been averse to searching for the living past on islands or precipice-guarded plateaus. Jefferson repeated the story of a trapper who had heard the mammoth roaring in the Virginia woods; in 1823 a South American traveler imaginatively viewed through his spyglass mastodons grazing in remote Andean valleys.

Nevertheless when the explorers had penetrated the last woodland, gazed on the last new animal—something they had pretty well accomplished by the middle of the nineteenth century—the past had been nowhere found. Only the great waters remained, the planetary expanse that, since the days of Thorfinn the Skull Cleaver, has received and, on occasion, swallowed the restless sails of men. Its surface was known, but its depths remained unplumbed. The treasure of countless piracies, the dead of innumerable battles had gone down into the green gloom of the mermaids' kingdom. In return, men had had only the legendary glimpse of a white arm at evening or the voice of a siren singing from some isle that would be gone at daybreak. Later, as men's youthful imaginations faded, only the rumor of sea monsters—serpents or archaic water beasts— survived from the abyss.

From a belief that the great deeps were lifeless,

scholars examining the growths on submarine cables and the scrapings brought up by newly devised dredges began to visualize something like Conan Doyle's *Lost World* in reverse. By 1870 this conception had two aspects: first, a theory that the ocean depths were populated by the living marine fossils of past geological ages which had here escaped the disasters that had destroyed their kind in the shallow seas of the earlier world; second—and reflecting the materialistic philosophy which was beginning to arise under the stimulus of the Darwinian theory—a belief that widespread on the floor of the abyssal plain lay the "*Urschleim*," a protoplasmic half-living matter representing that transition between the living and the nonliving out of which more complex life had, in the course of time, developed. The abyss, in other words, was thought to contain not only the living record of the past, but the ultimate secret of life itself; Creation might still be in process. Sir Charles Thomson in one enthusiastic statement in his *Depths of the Sea* even ventured to maintain: "The [depth] range of the various groups in modern seas corresponds remarkably with their vertical range in ancient strata." Down at the bottom, of course, lay that living undifferentiated primordial ooze as deep in the sea as it lay deep in time.

As the number of deep-sea soundings increased, as men slowly grasped the antiquity of that dark, cold world that is called the abyssal plain, a new idea arose:

the notion, as I have hinted, of a lost world in reverse, a midnight city of refuge in which the present mingled and lived on with the past. It was, of course, the world of the uttermost depths, the place without light since the beginning, and whose extent no continent above the waters could ever fill.

Of all the worlds of life the abyss alone remains unaltered. It is the one place on the planet where conditions remain as they have been since the beginning, where the five-mile pressures have not altered, where no suns have ever shone, where the cold is the same at the poles as at the equator, where the seasons are unchanging, where there is no wind and no wave to stir the ooze above which the glass sponges rise on graceful stems, or the abyssal sea squirts float like little balloons on strings above the mud. This is the sole world on the planet which we can enter only by a great act of the imagination. There has been, perhaps, only one greater imaginative effort—the attempt of nineteenth-century biology, intoxicated by its own successes, to observe on the sea floor life in the process of becoming, to glimpse in the abyssal oozes the crossing between life and death.

The story begins with the laying of the first Atlantic cable in the sixties of the last century. It involves one of the most peculiar and fantastic errors ever committed in the name of science. It is useless to blame this error upon one man because many leading figures of the day participated in what was, and remains, one of the most curious cases of self-delusion ever indulged in by scholars. It was the product of an overconfident materialism, a vainglorious assumption that the secrets of life were about to be revealed.

Haeckel in Germany and Huxley in England were proceeding to show that as one passed below the stage of nucleated single-celled organisms one arrived at a simple stirring of the abyssal slime wherein something that was neither life nor non-life oozed and fed without cellular individuality.

This soft, gelatinous matter had been taken from the ocean bed during dredging operations. Examined and pronounced upon by Professor Huxley, it was given the name of *Bathybius haeckelii* in honor of his great German colleague. Speaking before the Royal Geographical Society in 1870, Huxley confidently maintained that *Bathybius* formed a living scum or film on the sea bed extending over thousands of square miles. Moreover, he expanded, it probably formed a continuous sheet of living matter girdling the whole surface of the earth.

Sir Charles Thomson shared this view, commenting that the "organism" showed "no trace of differentiation of organs" and consisted apparently "of an amorphous sheet of a protein compound, irritable to a low degree and capable of assimilating food . . . a diffused formless protoplasm." Haeckel conceived of these formless "monera" as arising from non-living matter, their vital phenomena being traceable to "physico-chemical causes." Here was the "*Urschleim*" with a vengeance, the seething, unindividualized ooze whose potentialities included the butterfly and the rose. Man was mud and mud was man. Mechanism was the order of the day.

Unfortunately for this beautiful theory wistfully remembered by one writer as "explaining so much," *Bathybius* proved to be what the microscopists call an artifact; that is, it did not exist. A certain unfeeling Mr. Buchanan of the *Challenger* Expedition discovered, as he tried to investigate the nature of *Bathybius*, that he could produce all the characters of that indescribable animal by the simple process of adding strong alcohol to sea water. It was not necessary to drink the potion. One simply examined a specimen under the lens and observed that sulphate of lime was precipitated in the form of a gelatinous ooze which clung around particles as though ingesting them, thus lending a superficial protoplasmic appearance to the solution.

Mr. Huxley's original specimen had apparently been treated in this manner when it was sent to him. Huxley took the episode in good grace, but it was a severe blow to the materialists. The structureless protoplasmic "*Urschleim*" was a projective dream of scientists striving to build an evolutionary family tree upon existing organisms. Being nineteenth-century zoologists they unfortunately forgot the world of microscopic plant life, its basic position in the nourishment of living things, and the fact that it must have sunlight in order to perform its mysterious green miracles.

The abyss, it was now to be learned, whatever might roam its waters or slither wetly through its midnights, was not the original abode of life. If there was a past on the black plain far beneath us, if indeed the strange life of remote eras lingered there, it was not stacked with the layered neatness of geological strata as some oceanographers had imagined. The floating heads with their starveling bodies, the squid which emitted clouds of luminescent ink and vanished in their own bright explosions, were all a part of one of life's strangest qualities—its eternal dissatisfaction with what is, its persistent habit of reaching out into new environments and, by degrees, adapting itself to the most fantastic circumstances.

Once long ago as a child I can remember removing the cover from an old well. I was alone at the time and I can still anticipate, with a slight crawling of my scalp, the sight I inadvertently saw as I peered over the brink and followed a shaft of sunlight many feet down into the darkness. It touched, just touched in passing, a rusty pipe which projected across the well space some twenty feet above the water. And there, secretive as that very underground whose mystery had lured me into this adventure, I saw, passing surely and unhurriedly into the darkness, a spidery thing of hair and many legs. I set the rotting cover of boards back into place with a shiver, but that unidentifiable creature of the well has stayed with me to this day.

For the first time I must have realized, I think, the frightening diversity of the living; something that did not love the sun was down there, something that could walk through total darkness upon slender footholds over evil waters, something that had come down there by preference from above. It was in this way that the oceanic abyss was entered: by preference from above. Life did not arise on the bottom; the muds of the deep waters did not compound it. Instead, with its own pale lanterns or with the delicate, strawlike feelers of blindness, it has groped its way down into the dark.

The four-year voyage of the *Challenger* under the auspices of the British Admiralty, beginning in 1872, was the most ambitious project to investigate the ocean depths that men had ever attempted. The vessel was equipped with floating laboratories and a staff of naturalists. She traveled sixty-nine thousand nautical miles, took hundreds of soundings, and the observations of her staff of investigators occupy fifty huge volumes.

When the *Challenger* left port oceanography was still essentially a speculative science. Her biological director, Sir Charles Wyville Thomson, the same zoologist who had dredged the little red sea urchin out of the North Atlantic, believed along with many of his colleagues that the deep recesses of the ocean, unchanging through the ages, would reveal "living fossils," actual missing links in the history of life. Thomas Huxley, then at the height of his powers, proclaimed with characteristic vigor:

It may be confidently assumed that . . . the things brought up will . . . be zoological antiquities which in the tranquil and little changed depths of the ocean have escaped the causes of destruction at work in the shallows and represent the predominant population of a past age.

This view was enthusiastically shared by the great Swiss naturalist, Louis Agassiz, who contended that in deep waters "we should expect to find representatives of earlier geological periods." Agassiz went even further and observed that it was the deep waters which today most closely approximated the conditions under which life had originally emerged. It was, he said, the depths of the ocean alone which could place animals under a pressure such as he believed corresponded to the heavy atmosphere of a young world.

These were the excited dreams of science in 1872 as the *Challenger* steamed out of port. Sixty-nine thousand miles and four years later her weary scientists came home. They had rocked sickeningly in all seas, had dragged with cumbersome and ill-devised apparatus the very bowels of Creation. They had handled rare forms of life, looked on things denied to ordinary men, and, above all, they had laid the foundations of a true science of the sea. Nevertheless, their eyes were empty.

The great globe-girdling carpet of the living ooze was gone—that evolutionary base in which the German scholars had seen "an infinite capacity for improvement in every conceivable direction." "Our ardor," wearily confessed Moseley, the coral specialist, "abated somewhat . . . as the same tedious animals kept appearing from the depths in all parts of the world."

In the beginning even the cabin boys had crowded to see what four miles of rope would bring up from the bottom. Gradually, however, as the novelty wore off, the spectators became fewer. Even members of the scientific staff were not always present, particularly when the dredge arrived during the dinner hour.

The great hopes of the beginning were fading in disappointment, but Moseley gives an unforgettable picture of Sir Charles Thomson's sturdy persistence and enthusiasm in the face of the collapse of his theories. "To the last," he writes, "every cuttlefish which came up in our deep sea net was squeezed to see if it had a belemnite's bone in its back, and trilobites were eagerly looked out for." Either of these events would have found the world of the Paleozoic floundering alive on the deck of the *Challenger*. To the despair of Sir Charles they never appeared. It is true that here and there a few animals were recovered that were believed to be extinct and to exist only as fossils, but these were only such discoveries as might be expected when any vast unexplored region is first investigated, whether it be land or sea.

The secret and remote abysses were yielding not the protected remnants of the very earliest world, but a scattering of later antique types along with a more modern abyssal fauna obviously related to, and descended from, the swarming creatures of the shallow seas and upper waters. Such ancient forms as survive in the abyss represent adaptations and migrations that took place in antiquity from the continental shoals far above. In that sense the midnight timeless city does indeed exist, for in those depths the ages overlap and some few elements of the older world, losing out in competition with more highly evolved and modern types, have chosen to slip by degrees into the freezing cold of the abyss. Here in the unchanging mud and comforting darkness they have survived. After them in time have come others, groping into that enormous cellar with lanterns or light-magnifying eyes—clever adaptations possible to squids and higher vertebrates.

Even among the mammals, the great sperm whale has come sounding down into the fearful pressures of the kraken's world, the last of all to enter, and capable of enduring only moments on what is actually the upper edge of the abyss. If it is a place of refuge it is also, we know now, a famine world. There is no vegetable life below there. All that lives preys on others or on the dead raining down from above. This is the reason for the curiously abbreviated bodies of many of the fishes and their enormous jaws; this is the reason why we know that life came relatively late to the abyss.

According to the biochemists the conditions under which cellular life is possible are very restricted, nor have they changed in any marked degree since life began. At first glance this statement seems absurd. Life has crept upward from the waters, it crawls in the fields, it penetrates the air, it is not unknown even in the frozen wastes of the Antarctic. Surely this enormous diversity is the very reverse of restriction.

The answer, of course, lies in that modest little phrase "the conditions of cellular life." All of the tremendous differences between living forms have been achieved only by the elaboration of devices for the maintenance of that inner nourishing liquidity in which cells can live and grow within a certain narrow range of tolerance. Not for nothing has the composition of mammalian blood led to our description as "walking sacks of sea water." Not for nothing did the great French physiologist Bernard comment that "the stability of the interior environment is the condition of free life."

The drifting cell masses of the early ocean lived in a

nutrient solution. Salt and sun and moisture were accessible without great mechanical elaboration. It was the reaching out that changed this pattern, the reaching out that forced the cells to bring the sea ashore with them, to elaborate in their own bodies the very miniature of that all-embracing sea from which they came. It was the reaching out, that magnificent and agelong groping that only life—blindly and persistently among stones and the indifference of the entire inanimate universe—can continue to endure and prolong.

Men have worked in many places. They have seen this sea-born protoplasm creeping upward in the shape of lichens, among the howling winds of snow-clad mountains. They have seen it in the delicate "snowshoe" feet of desert lizards devised for running over sand. From some unknown spot, most probably along the shoals above the continental shelf, it has reached out into lakes and grasslands, edged stealthily into deserts, learned even to endure the heat of boiling springs or to hatch eggs, like the emperor penguin, in the blizzards by the southern pole. It has similarly found its way into the downward coursing streams of the abyss. It has solved the pressures of the ocean bottom as it has survived the rarefied air of the highest mountains. In these difficult surroundings life thins a little; the inventions that support it grow more difficult to produce and the intrusions are apt to be late, because life has experimented last in these bleak planetary wastelands.

Nevertheless the reaching out that began a billion years ago is still in process. The cells, so carefully transferring their limited range of endurance through astounding extremes of heat and frost and pressure, show no inclination toward content. Content is a word unknown to life; it is also a word unknown to man.

In 1949, on the White Sands proving grounds, a Wac Corporal rocket reached an altitude of 250 miles and, on the verge of outer space, paused and fell back. Somehow I like to think of those rockets, pounding year after year at that ocean of air, roaring away into an immensity from which, before long, one will not come back. Sometimes, walking in the star-sprinkled evenings, I think of that almost forgotten theory of Arrhenius that the spores of life came originally from outer space.

Perhaps that explains it, I think wistfully—life reaching out, groping for a billion years, life desperate to go home.

The nineteenth-century mechanists, at least, did not find our origins in the abyss, and every bubble of the chemist's broth has left the secret of life as inscrutably remote as ever. The ingredients are known; they are to be had on any drug-store shelf. You can take them yourself and pour them and and wait hopefully for the resulting slime to crawl. It will not. The beautiful pulse of streaming protoplasm, that unknown organization of an unstable chemistry which makes up the life process, will not begin. Carbon, nitrogen, hydrogen, and oxygen you have mixed, and the same dead chemicals they remain.

Shape of sea water and carbon rings, yet simultaneously a perplexed professor on a village street, I look up across the moon and Venus—outward, outward into that blue-white glitter beyond the galaxy. And as I look and shiver I feel the voice in every fiber of my being: Have we come from elsewhere? By these our instruments shall we go home? Whatever the beginning, and by whatever mechanical extensions, life is about to cross into the open domain of space. Has not the great 200-inch reflector upon Mount Palomar already spied out the prospect?

A billion years have gone into the making of that eye; the water and the salt and the vapors of the sun have built it; things that squirmed in the tide silts have devised it. Light-year beyond light-year, deep beyond deep, the mind may rove by means of it, hanging above the bottomless and surveying impartially the state of matter in the white-dwarf suns.

Yet whenever I see a frog's eye low in the water warily ogling the shoreward landscape, I always think inconsequentially of those twiddling mechanical eyes that mankind manipulates nightly from a thousand observatories. Someday, with a telescopic lens an acre in extent, we are going to see something not to our liking, some looming shape outside there across the great pond of space.

Whenever I catch a frog's eye I am aware of this, but I do not find it depressing. I stand quite still and try hard not to move or lift a hand since it would only frighten him. And standing thus it finally comes to me that this is the most enormous extension of vision of which life is capable: the projection of itself into other lives. This is the lonely, magnificent power of humanity. It is, far more than any spatial adventure, the supreme epitome of the reaching out.

THE SNOUT

🌳

I have long been an admirer of the octopus. The cephalopods are very old, and they have slipped, protean, through many shapes. They are the wisest of the mollusks, and I have always felt it to be just as well for us that they never came ashore, but—there are other things that have.

There is no need to be frightened. It is true some of the creatures are odd, but I find the situation rather heartening than otherwise. It gives one a feeling of confidence to see nature still busy with experiments, still dynamic, and not through nor satisfied because a Devonian fish managed to end as a two-legged character with a straw hat. There are other things brewing and growing in the oceanic vat. It pays to know this. It pays to know there is just as much future as there is past. The only thing that doesn't pay is to be sure of man's own part in it.

There are things down there still coming ashore. Never make the mistake of thinking life is now adjusted for eternity. It gets into your head—the certainty, I mean—the human certainty, and then you miss it all: the things on the tide flats and what they mean, and why, as my wife says, "they ought to be watched."

The trouble is we don't know what to watch for. I have a friend, one of these Explorers Club people, who drops in now and then between trips to tell me about the size of crocodile jaws in Uganda, or what happened on some back beach in Arnhem Land.

"They fell out of the trees," he said. "Like rain. And into the boat."

"Uh?" I said, noncommittally.

"They did *so*," he protested, "and they were hard to catch."

"Really—" I said.

"We were pushing a dugout up one of the tidal creeks in northern Australia and going fast when *smacko* we jam this mangrove bush and the things come tumbling down.

"What were they doing sitting up there in bunches? I ask you. It's no place for a fish. Besides that they had a way of sidling off with those popeyes trained on you. I never liked it. Somebody ought to keep an eye on them."

"Why?" I asked.

"I don't know why," he said impatiently, running a rough, square hand through his hair and wrinkling his forehead. "I just mean they make you feel that way, is all. A fish belongs in the water. It ought to stay there —just as we live on land in houses. Things ought to know their place and stay in it, but those fish have got a way of sidling off. As though they had mental reservations and weren't keeping any contracts. See what I mean?"

"I see what you mean," I said gravely. "They ought to be watched. My wife thinks so too. About a lot of things."

"She does?" He brightened. "Then that's two of us. I don't know why, but they give you that feeling."

He didn't know why, but I thought that I did.

It began as such things always begin—in the ooze of unnoticed swamps, in the darkness of eclipsed moons. It began with a strangled gasping for air.

The pond was a place of reek and corruption, of fetid smells and of oxygen-starved fish breathing through laboring gills. At times the slowly contracting circle of the water left little windrows of minnows who skittered desperately to escape the sun, but who died, nevertheless, in the fat, warm mud. It was a place of low life. In it the human brain began.

There were strange snouts in those waters, strange barbels nuzzling the bottom ooze, and there was time—three hundred million years of it—but mostly, I think, it was the ooze. By day the temperature in the world outside the pond rose to a frightful intensity; at night the sun went down in smoking red. Dust storms marched in incessant progression across a wilderness whose plants were the plants of long ago. Leafless and weird and stiff they lingered by the water, while over vast areas of grassless uplands the winds blew until red stones took on the polish of reflecting mirrors. There was nothing to hold the land in place. Winds howled, dust clouds rolled, and brief erratic torrents choked with silt ran down to the sea. It was a time of dizzying contrasts, a time of change.

On the oily surface of the pond, from time to time a snout thrust upward, took in air with a queer grunting inspiration, and swirled back to the bottom. The pond was doomed, the water was foul, and the oxygen almost gone, but the creature would not die. It could breathe air direct through a little accessory lung, and it could walk. In all that weird and lifeless landscape, it was the only thing that could. It walked rarely and under protest, but that was not surprising. The creature was a fish.

In the passage of days the pond became a puddle, but

the Snout survived. There was dew one dark night and a coolness in the empty stream bed. When the sun rose next morning the pond was an empty place of cracked mud, but the Snout did not lie there. He had gone. Down stream there were other ponds. He breathed air for a few hours and hobbled slowly along on the stumps of heavy fins.

It was an uncanny business if there had been anyone there to see. It was a journey best not observed in daylight, it was something that needed swamps and shadows and the touch of the night dew. It was a monstrous penetration of a forbidden element, and the Snout kept his face from the light. It was just as well, though the face should not be mocked. In three hundred million years it would be our own.

There was something fermenting in the brain of the Snout. He was no longer entirely a fish. The ooze had marked him. It takes a swamp-and-tide-flat zoologist to tell you about life; it is in this domain that the living suffer great extremes, it is here that the water-failures, driven to desperation, make starts in a new element. It is here that strange compromises are made and new senses are born. The Snout was no exception. Though he breathed and walked primarily in order to stay in the water, he was coming ashore.

He was not really a successful fish except that he was managing to stay alive in a noisome, uncomfortable, oxygen-starved environment. In fact the time was coming when the last of his kind, harried by more ferocious and speedier fishes, would slip off the edge of the continental shelf, to seek safety in the sunless abysses of the deep sea. But the Snout was a fresh-water Crossopterygian, to give him his true name, and cumbersome and plodding though he was, something had happened back of his eyes. The ooze had gotten in its work.

It is interesting to consider what sort of creatures we, the remote descendants of the Snout, might be, except for that green quagmire out of which he came. Mammalian insects perhaps we should have been—solid-brained, our neurones wired for mechanical responses, our lives running out with the perfection of beautiful, intricate, and mindless clocks. More likely we should never have existed at all. It was the Snout and the ooze that did it. Perhaps there also, among rotting fish heads and blue, night-burning bog lights, moved the eternal mystery, the careful finger of God. The increase was not much. It was two bubbles, two thin-walled little balloons at the end of the Snout's small brain. The cerebral hemispheres had appeared.

Among all the experiments in that dripping, ooze-filled world, one was vital: the brain had to be fed. The nerve tissues are insatiable devourers of oxygen. If they do not get it, life is gone. In stagnant swamp waters, only the development of a highly efficient blood supply to the brain can prevent disaster. And among those gasping, dying creatures, whose small brains winked out forever in the long Silurian drought, the Snout and his brethren survived.

Over the exterior surface of the Snout's tiny brain ran the myriad blood vessels that served it; through the greatly enlarged choroid plexuses, other vessels pumped oxygen into the spinal fluid. The brain was a thin-walled tube fed from both surfaces. It could only exist as a thing of thin walls permeated with oxygen. To thicken, to lay down solid masses of nervous tissue such as exist among the fishes in oxygenated waters was to invite disaster. The Snout lived on a bubble, two bubbles in his brain.

It was not that his thinking was deep; it was only that it had to be thin. The little bubbles of the hemispheres helped to spread the area upon which higher correlation centers could be built, and yet preserve those areas from the disastrous thickenings which meant oxygen death to the swamp dweller. There is a mystery about those thickenings which culminate in the so-called solid brain. It is the brain of insects, of the modern fishes, of some reptiles and all birds. Always it marks the appearance of elaborate patterns of instinct and the end of thought. A road has been taken which, anatomically, is well-nigh irretraceable; it does not lead in the direction of a high order of consciousness.

Wherever, instead, the thin sheets of gray matter expand upward into the enormous hemispheres of the human brain, laughter, or it may be sorrow, enters in. Out of the choked Devonian waters emerged sight and sound and the music that rolls invisible through the composer's brain. They are there still in the ooze along the tideline, though no one notices. The world is fixed, we say: fish in the sea, birds in the air. But in the mangrove swamps by the Niger, fish climb trees and ogle uneasy naturalists who try unsuccessfully to chase them back to the water. There are things still coming ashore.

The door to the past is a strange door. It swings open and things pass through it, but they pass in one direction only. No man can return across that threshold, though he can look down still and see the green light waver in the water weeds.

There are two ways to seek the doorway: in the swamps of the inland waterways and along the tide flats of the estuaries where rivers come to the sea. By those two pathways life came ashore. It was not the magnificent march through the breakers and up the cliffs that we fondly imagine. It was a stealthy advance made in suffocation and terror, amidst the leaching bite of chemical discomfort. It was made by the failures of the sea.

Some creatures have slipped through the invisible chemical barrier between salt and fresh water into the tidal rivers, and later come ashore; some have crept upward from the salt. In all cases, however, the first adventure into the dreaded atmosphere seems to have been largely determined by the inexorable crowding of enemies and by the retreat further and further into marginal situations where the oxygen supply was depleted. Finally, in the ruthless selection of the swamp margins, or in the scramble for food on the tide flats, the land becomes home.

Not the least interesting feature of some of the tide-flat emergents is their definite antipathy for the full tide. It obstructs their food-collecting on the mud banks and brings their enemies. Only extremes of fright will drive them into the water for any period.

I think it was the great nineteenth-century paleontol·ogist Cope who first clearly enunciated what he called the "law of the unspecialized," the contention that it was not from the most highly organized and dominant forms of a given geological era that the master type of a succeeding period evolved, but that instead the dominant forms tended to arise from more lowly and generalized animals which were capable of making new adaptations, and which were not narrowly restricted to a given environment.

There is considerable truth to this observation, but, for all that, the idea is not simple. Who is to say without foreknowledge of the future which animal is specialized and which is not? We have only to consider our remote ancestor, the Snout, to see the intricacies into which the law of the unspecialized may lead us.

If we had been making zoological observations in the Paleozoic Age, with no knowledge of the strange realms life was to penetrate in the future, we would probably have regarded the Snout as specialized. We would have seen his air-bladder lung, his stubby, sluggish fins, and his odd ability to wriggle overland as specialized adaptations to a peculiarly restricted environmental niche in stagnant continental waters. We would have thought in water terms and we would have dismissed the Snout as an interesting failure off the main line of progressive evolution, escaping from his enemies and surviving successfully only in the dreary and marginal surroundings scorned by the swift-finned teleost fishes who were destined to dominate the seas and all quick waters.

Yet it was this poor specialization—this bog-trapped failure—whose descendants, in three great movements, were to dominate the earth. It is only now, looking backward, that we dare to regard him as "generalized." The Snout was the first vertebrate to pop completely through the water membrane into a new dimension. His very specializations and failures, in a water sense, had preadapted him for a world he scarcely knew existed.

The day of the Snout was over three hundred million years ago. Not long since I read a book in which a prominent scientist spoke cheerfully of some ten billion years of future time remaining to us. He pointed out happily the things that man might do throughout that period. Fish in the sea, I thought again, birds in the air. The climb all far behind us, the species fixed and sure. No wonder my explorer friend had had a momentary qualm when he met the mudskippers with their mental reservations and lack of promises. There is something wrong with our world view. It is still Ptolemaic, though the sun is no longer believed to revolve around the earth.

We teach the past, we see farther backward into time than any race before us, but we stop at the present, or, at best, we project far into the future idealized versions of ourselves. All that long way behind us we see, perhaps inevitably, through human eyes alone. We see ourselves as the culmination and the end, and if we do indeed consider our passing, we think that sunlight will go with us and the earth be dark. We are the end. For us continents rose and fell, for us the waters and the air were mastered, for us the great living web has pulsated and grown more intricate.

To deny this, a man once told me, is to deny God. This puzzled me. I went back along the pathway to the marsh. I went, not in the past, not by the bones of dead things, not down the lost roadway of the Snout. I went instead in daylight, in the Now, to see if the door was still there, and to see what things passed through.

I found that the same experiments were brewing, that up out of that ancient well, fins were still scrambling toward the sunlight. They were small things, and which of them presaged the future I could not say. I saw only that they were many and that they had solved the oxygen death in many marvelous ways, not always ours.

I found that there were modern fishes who breathed air, not through a lung but through their stomachs or through strange chambers where their gills should be, or breathing as the Snout once breathed. I found that some crawled in the fields at nightfall pursuing insects, or slept on the grass by pond sides and who drowned, if kept under water, as men themselves might drown.

Of all these fishes the mudskipper *Periophthalmus* is perhaps the strangest. He climbs trees with his fins and pursues insects; he snaps worms like a robin on the tide flats; he sees as land things see, and above all he dodges and evades with a curious popeyed insolence more suggestive of the land than of the sea. Of a different tribe and a different time he is, nevertheless, oddly reminiscent of the Snout.

But not the same. There lies the hope of life. The old ways are exploited and remain, but new things come, new senses try the unfamiliar air. There are small scuttlings and splashings in the dark, and out of it come the first croaking, illiterate voices of the things to be, just as man once croaked and dreamed darkly in that tiny vesicular forebrain.

Perpetually, now, we search and bicker and disagree. The eternal form eludes us—the shape we conceive as ours. Perhaps the old road through the marsh should tell us. We are one of many appearances of the thing called Life; we are not its perfect image, for it has no image except Life, and life is multitudinous and emergent in the stream of time.

HOW FLOWERS
CHANGED THE WORLD

If it had been possible to observe the Earth from the far side of the solar system over the long course of geological epochs, the watchers might have been able to discern a subtle change in the light emanating from our planet. That world of long ago would, like the red deserts of Mars, have reflected light from vast drifts of stone and gravel, the sands of wandering wastes, the blackness of naked basalt, the yellow dust of endlessly moving storms. Only the ceaseless marching of the clouds and the intermittent flashes from the restless surface of the sea would have told a different story, but still essentially a barren one. Then, as the millennia rolled away and age followed age, a new and greener light would, by degrees, have come to twinkle across those endless miles.

This is the only difference those far watchers, by the use of subtle instruments, might have perceived in the whole history of the planet Earth. Yet that slowly growing green twinkle would have contained the epic march of life from the tidal oozes upward across the raw and unclothed continents. Out of the vast chemical bath of the sea—not from the deeps, but from the element-rich, light-exposed platforms of the continental shelves —wandering fingers of green had crept upward along the meanderings of river systems and fringed the gravels of forgotten lakes.

In those first ages plants clung of necessity to swamps and watercourses. Their reproductive processes demanded direct access to water. Beyond the primitive ferns and mosses that enclosed the borders of swamps and streams the rocks still lay vast and bare, the winds still swirled the dust of a naked planet. The grass cover that holds our world secure in place was still millions of years in the future. The green marchers had gained a soggy foothold upon the land, but that was all. They did not reproduce by seeds but by microscopic swimming sperm that had to wriggle their way through water to fertilize the female cell. Such plants in their higher forms had clever adaptations for the use of rain water in their sexual phases, and survived with increasing success in a wet land environment. They now seem part of man's normal environment. The truth is, however, that there is nothing very "normal" about nature. Once upon a time there were no flowers at all.

A little while ago—about one hundred million years, as the geologist estimates time in the history of our four-billion-year-old planet—flowers were not to be found anywhere on the five continents. Wherever one might have looked, from the poles to the equator, one would have seen only the cold dark monotonous green of a world whose plant life possessed no other color.

Somewhere, just a short time before the close of the Age of Reptiles, there occurred a soundless, violent explosion. It lasted millions of years, but it was an explosion, nevertheless. It marked the emergence of the angiosperms—the flowering plants. Even the great evolutionist, Charles Darwin, called them "an abominable mystery," because they appeared so suddenly and spread so fast.

Flowers changed the face of the planet. Without them, the world we know—even man himself—would never have existed. Francis Thompson, the English poet, once wrote that one could not pluck a flower without troubling a star. Intuitively he had sensed like a naturalist the enormous interlinked complexity of life.

Today we know that the appearance of the flowers contained also the equally mystifying emergence of man.

If we were to go back into the Age of Reptiles, its drowned swamps and birdless forests would reveal to us a warmer but, on the whole, a sleepier world than that of today. Here and there, it is true, the serpent heads of bottom-feeding dinosaurs might be upreared in suspicion of their huge flesh-eating compatriots. Tyrannosaurs, enormous bipedal caricatures of men, would stalk mindlessly across the sites of future cities and go their slow way down into the dark of geologic time.

In all that world of living things nothing saw save with the intense concentration of the hunt, nothing moved except with the grave sleepwalking intentness of the instinct-driven brain. Judged by modern standards, it was a world in slow motion, a cold-blooded world whose occupants were most active at noonday but torpid on chill nights, their brains damped by a slower metabolism than any known to even the most primitive of warm-blooded animals today.

A high metabolic rate and the maintenance of a constant body temperature are supreme achievements in the evolution of life. They enable an animal to escape, within broad limits, from the overheating or the chilling of its immediate surroundings, and at the same time to maintain a peak mental efficiency. Creatures without a high metabolic rate are slaves to weather. Insects in the first frosts of autumn all run down like little clocks. Yet if you pick one up and breathe warmly upon it, it will begin to move about once more.

In a sheltered spot such creatures may sleep away the winter, but they are hopelessly immobilized. Though a few warm-blooded mammals, such as the woodchuck of our day, have evolved a way of reducing their metabolic rate in order to undergo winter hibernation, it is a survival mechanism with drawbacks, for it leaves the animal helplessly exposed if enemies discover him during his period of suspended animation. Thus bear or woodchuck, big animal or small, must seek, in this time of descending sleep, a safe refuge in some hidden den or burrow. Hibernation is, therefore, primarily a winter refuge of small, easily concealed animals rather than of large ones.

A high metabolic rate, however, means a heavy intake of energy in order to sustain body warmth and efficiency. It is for this reason that even some of these later warm-blooded mammals existing in our day have learned to descend into a slower, unconscious rate of living during the winter months when food may be difficult to obtain. On a slightly higher plane they are following the procedure of the cold-blooded frog sleeping in the mud at the bottom of a frozen pond.

The agile brain of the warm-blooded birds and mammals demands a high oxygen consumption and food in concentrated forms, or the creatures cannot long sustain themselves. It was the rise of the flowering plants that provided that energy and changed the nature of the living world. Their appearance parallels in a quite surprising manner the rise of the birds and mammals.

Slowly, toward the dawn of the Age of Reptiles, something over two hundred and fifty million years ago, the little naked sperm cells wriggling their way through dew and raindrops had given way to a kind of pollen carried by the wind. Our present-day pine forests represent plants of a pollen-disseminating variety. Once fertilization was no longer dependent on exterior water, the march over drier regions could be extended. Instead of spores simple primitive seeds carrying some nourishment for the young plant had developed, but true flowers were still scores of millions of years away. After a long period of hesitant evolutionary groping, they exploded upon the world with truly revolutionary violence.

The event occurred in Cretaceous times in the close of the Age of Reptiles. Before the coming of the flowering plants our own ancestral stock, the warm-blooded mammals, consisted of a few mousy little creatures hidden in trees and underbrush. A few lizard-like birds with carnivorous teeth flapped awkwardly on ill-aimed flights among archaic shrubbery. None of these insignificant creatures gave evidence of any remarkable talents. The mammals in particular had been around for some millions of years, but had remained well lost in the shadow of the mighty reptiles. Truth to tell, man was still, like the genie in the bottle, encased in the body of a creature about the size of a rat.

As for the birds, their reptilian cousins the Pterodactyls, flew farther and better. There was just one thing about the birds that paralleled the physiology of the mammals. They, too, had evolved warm blood and its accompanying temperature control. Nevertheless, if one had been seen stripped of his feathers, he would still have seemed a slightly uncanny and unsightly lizard.

Neither the birds nor the mammals, however, were quite what they seemed. They were waiting for the Age of Flowers. They were waiting for what flowers, and with them the true encased seed, would bring. Fish-eating, gigantic leather-winged reptiles, twenty-eight feet from wing tip to wing tip, hovered over the coasts that one day would be swarming with gulls.

Inland the monotonous green of the pine and spruce forests with their primitive wooden cone flowers stretched everywhere. No grass hindered the fall of the naked seeds to earth. Great sequoias towered to the skies. The world of that time has a certain appeal but it is a giant's world, a world moving slowly like the reptiles who stalked magnificently among the boles of its trees.

The trees themselves are ancient, slow-growing and immense, like the redwood groves that have survived to our day on the California coast. All is stiff, formal, upright and green, monotonously green. There is no grass as yet; there are no wide plains rolling in the sun, no tiny daisies dotting the meadows underfoot. There is little versatility about this scene; it is, in truth, a giant's world.

A few nights ago it was brought home vividly to me that the world has changed since that far epoch. I was awakened out of sleep by an unknown sound in my living room. Not a small sound—not a creaking timber or a mouse's scurry—but a sharp, rending explosion as though an unwary foot had been put down upon a wine glass. I had come instantly out of sleep and lay tense, unbreathing. I listened for another step. There was none.

Unable to stand the suspense any longer, I turned on the light and passed from room to room glancing uneasily behind chairs and into closets. Nothing seemed disturbed, and I stood puzzled in the center of the living room floor. Then a small button-shaped object upon the rug caught my eye. It was hard and polished and glistening. Scattered over the length of the room were several more shining up at me like wary little eyes. A pine cone that had been lying in a dish had been blown the length of the coffee table. The dish itself could hardly have been the source of the explosion. Beside it I found two ribbon-like strips of a velvety-green. I tried to place the two strips together to make a pod. They twisted resolutely away from each other and would no longer fit.

I relaxed in a chair, then, for I had reached a solution of the midnight disturbance. The twisted strips were wistaria pods that I had brought in a day or two previously and placed in the dish. They had chosen midnight to explode and distribute their multiplying fund of life down the length of the room. A plant, a fixed, rooted thing, immobilized in a single spot, had devised a way of propelling its offspring across open space. Immediately there passed before my eyes the million airy troopers of the milkweed pod and the clutching hooks of the sandburs. Seeds on the coyote's tail, seeds on the hunter's coat, thistledown mounting on the winds—all were somehow triumphing over life's limitations. Yet the ability to do this had not been with them at the beginning. It was the product of endless effort and experiment.

The seeds on my carpet were not going to lie stiffly where they had dropped like their antiquated cousins, the naked seeds on the pine-cone scales. They were travelers. Struck by the thought, I went out next day and collected several other varieties. I line them up now in a row on my desk—so many little capsules of life, winged, hooked or spiked. Every one is an angiosperm, a product of the true flowering plants. Contained in these little boxes is the secret of that far-off Cretaceous explosion of a hundred million years ago that changed the face of the planet. And somewhere in here, I think, as I poke seriously at one particularly resistant seedcase of a wild grass, was once man himself.

When the first simple flower bloomed on some raw upland late in the Dinosaur Age, it was wind pollinated, just like its early pine-cone relatives. It was a very inconspicuous flower because it had not yet evolved the idea of using the surer attraction of birds and insects to achieve the transportation of pollen. It sowed its own pollen and received the pollen of other flowers by the simple vagaries of the wind. Many plants in regions where insect life is scant still follow this principle today. Nevertheless, the true flower—and the seed that it produced—was a profound innovation in the world of life.

In a way, this event parallels, in the plant world, what happened among animals. Consider the relative chance for survival of the exteriorly deposited egg of a fish in contrast with the fertilized egg of a mammal, carefully retained for months in the mother's body until the young animal (or human being) is developed to a point where it may survive. The biological wastage is less—and so it is with the flowering plants. The primitive spore, a single cell fertilized in the beginning by a swimming sperm, did not promote rapid distribution, and the young plant, moreover, had to struggle up from nothing. No one had left it any food except what it could get by its own unaided efforts.

By contrast, the true flowering plants (angiosperm itself means "encased seed") grew a seed in the heart of a flower, a seed whose development was initiated by a fertilizing pollen grain independent of outside moisture. But the seed, unlike the developing spore, is already a fully equipped *embryonic plant* packed in a little enclosed box stuffed full of nutritious food. Moreover, by featherdown attachments, as in dandelion or milkweed seed, it can be wafted upward on gusts and ride the wind for miles; or with hooks it can cling to a bear's or a rabbit's hide; or like some of the berries, it can be covered with a juicy, attractive fruit to lure birds, pass undigested through their intestinal tracts and be voided miles away.

The ramifications of this biological invention were endless. Plants traveled as they had never traveled before. They got into strange environments heretofore never entered by the old spore plants or stiff pine-cone-seed plants. The well-fed, carefully cherished little embryos raised their heads everywhere. Many of the older plants with more primitive reproductive mechanisms began to fade away under this unequal contest. They contracted their range into secluded environments. Some, like the giant redwoods, lingered on as relics; many vanished entirely.

The world of the giants was a dying world. These fantastic little seeds skipping and hopping and flying about the woods and valleys brought with them an amazing adaptability. If our whole lives had not been spent in the midst of it, it would astound us. The old, stiff, sky-reaching wooden world had changed into something that glowed here and there with strange colors, put out queer, unheard-of fruits and little intricately carved seed cases, and, most important of all, produced concentrated foods in a way that the land had never seen before, or dreamed of back in the fish-eating, leaf-crunching days of the dinosaurs.

That food came from three sources, all produced by the reproductive system of the flowering plants. There were the tantalizing nectars and pollens intended to draw insects for pollenizing purposes, and which are responsible also for that wonderful jeweled creation, the hummingbird. There were the juicy and enticing fruits to attract larger animals, and in which tough-coated seeds were concealed, as in the tomato, for example. Then, as if this were not enough, there was the food in the actual seed itself, the food intended to nourish the embryo. All over the world, like hot corn in a popper, these incredible elaborations of the flowering plants kept exploding. In a movement that was almost instantaneous, geologically speaking, the angiosperms had taken over the world. Grass was beginning to cover the bare earth until, today, there are over six thousand species. All kinds of vines and bushes squirmed and writhed under new trees with flying seeds.

The explosion was having its effect on animal life also. Specialized groups of insects were arising to feed on the new sources of food and, incidentally and unknowingly, to pollinate the plant. The flowers bloomed and bloomed in ever larger and more spectacular varieties. Some were pale unearthly night flowers intended to lure moths in the evening twilight, some among the orchids even took the shape of female spiders in order to attract wandering males, some flamed redly in the light of noon or twinkled modestly in the meadow grasses. Intricate mechanisms splashed pollen on the breasts of hummingbirds, or stamped it on the bellies of black, grumbling bees droning assiduously from blossom to blossom. Honey ran, insects multiplied, and even the descendants of that toothed and ancient lizard-bird had become strangely altered. Equipped with prodding beaks instead of biting teeth they pecked the seeds and gobbled the insects that were really converted nectar.

Across the planet grasslands were now spreading. A slow continental upthrust which had been a part of the early Age of Flowers had cooled the world's climates. The stalking reptiles and the leather-winged black imps of the seashore cliffs had vanished. Only birds roamed the air now, hot-blooded and high-speed metabolic machines.

The mammals, too, had survived and were venturing

into new domains, staring about perhaps a bit bewildered at their sudden eminence now that the thunder lizards were gone. Many of them, beginning as small browsers upon leaves in the forest, began to venture out upon this new sunlit world of the grass. Grass has a high silica content and demands a new type of very tough and resistant tooth enamel, but the seeds taken incidentally in the cropping of the grass are highly nutritious. A new world had opened out for the warm-blooded mammals. Great herbivores like the mammoths, horses and bisons appeared. Skulking about them had arisen savage flesh-feeding carnivores like the now extinct dire wolves and the saber-toothed tiger.

Flesh eaters though these creatures were, they were being sustained on nutritious grasses one step removed. Their fierce energy was being maintained on a high, effective level, through hot days and frosty nights, by the concentrated energy of the angiosperms. That energy, thirty per cent or more of the weight of the entire plant among some of the cereal grasses, was being accumulated and concentrated in the rich proteins and fats of the enormous game herds of the grasslands.

On the edge of the forest, a strange, old-fashioned animal still hesitated. His body was the body of a tree dweller, and though tough and knotty by human standards, he was, in terms of that world into which he gazed, a weakling. His teeth, though strong for chewing on the tough fruits of the forest, or for crunching an occasional unwary bird caught with his prehensile hands, were not the tearing sabers of the great cats. He had a passion for lifting himself up to see about, in his restless, roving curiosity. He would run a little stiffly and uncertainly, perhaps, on his hind legs, but only in those rare moments when he ventured out upon the ground. All this was the legacy of his climbing days; he had a hand with flexible fingers and no fine specialized hoofs upon which to gallop like the wind.

If he had any idea of competing in that new world, he had better forget it; teeth or hooves, he was much too late for either. He was a ne'er-do-well, an in-betweener. Nature had not done well by him. It was as if she had hesitated and never quite made up her mind. Perhaps as a consequence he had a malicious gleam in his eye, the gleam of an outcast who has been left nothing and knows he is going to have to take what he gets. One day a little band of these odd apes—for apes they were—shambled out upon the grass; the human story had begun.

Apes were to become men, in the inscrutable wisdom of nature, because flowers had produced seeds and fruits in such tremendous quantities that a new and totally different store of energy had become available in concentrated form. Impressive as the slow-moving, dim-brained dinosaurs had been, it is doubtful if their age had supported anything like the diversity of life that now rioted across the planet or flashed in and out among the trees. Down on the grass by a streamside, one of those apes with inquisitive fingers turned over a stone and hefted it vaguely. The group clucked together in a throaty tongue and moved off through the tall grass foraging for seeds and insects. The one still held, sniffed, and hefted the stone he had found. He liked the feel of it in his fingers. The attack on the animal world was about to begin.

If one could run the story of that first human group like a speeded-up motion picture through a million years of time, one might see the stone in the hand change to the flint ax and the torch. All that swarming grassland world with its giant bison and trumpeting mammoths would go down in ruin to feed the insatiable and growing numbers of a carnivore who, like the great cats before him, was taking his energy indirectly from the grass. Later he found fire and it altered the tough meats and drained their energy even faster into a stomach ill adapted for the ferocious turn man's habits had taken.

His limbs grew longer, he strode more purposefully over the grass. The stolen energy that would take man across the continents would fail him at last. The great Ice Age herds were destined to vanish. When they did so, another hand like the hand that grasped the stone by the river long ago would pluck a handful of grass seed and hold it contemplatively.

In that moment, the golden towers of man, his swarming millions, his turning wheels, the vast learning of his packed libraries, would glimmer dimly there in the ancestor of wheat, a few seeds held in a muddy hand. Without the gift of flowers and the infinite diversity of their fruits, man and bird, if they had continued to exist at all, would be today unrecognizable. Archaeopteryx, the lizard-bird, might still be snapping at beetles on a sequoia limb; man might still be a nocturnal insectivore gnawing a roach in the dark. The weight of a petal has changed the face of the world and made it ours.

THE REAL SECRET
OF PILTDOWN

🌳

How did man get his brain? Many years ago Charles Darwin's great contemporary, and co-discoverer with him of the principle of natural selection, Alfred Russel Wallace, propounded that simple question. It is a question which has bothered evolutionists ever since, and when Darwin received his copy of an article Wallace had written on this subject he was obviously shaken. It is recorded that he wrote in anguish across the paper, "No!" and underlined the "No" three times heavily in a rising fervor of objection.

Today the question asked by Wallace and never satisfactorily answered by Darwin has returned to haunt us.

A skull, a supposedly very ancient skull, long used as one of the most powerful pieces of evidence documenting the Darwinian position upon human evolution, has been proven to be a forgery, a hoax perpetrated by an unscrupulous but learned amateur. In the fall of 1953 the famous Piltdown cranium, known in scientific circles all over the world since its discovery in a gravel pit on the Sussex Downs in 1911, was jocularly dismissed by the world's press as the skull that had "made monkeys out of the anthropologists." Nobody remembered in 1953 that Wallace, the great evolutionist, had protested to a friend in 1913, "The Piltdown skull does not prove much, if anything!"

Why had Wallace made that remark? Why, almost alone among the English scientists of his time, had he chosen to regard with a dubious eye a fossil specimen that seemed to substantiate the theory to which he and Darwin had devoted their lives? He did so for one reason: he did not believe what the Piltdown skull appeared to reveal as to the nature of the process by which the human brain had been evolved. He did not believe in a skull which had a modern brain box attached to an apparently primitive face and given, in the original estimates, an antiquity of something over a million years.

Today we know that the elimination of the Piltdown skull from the growing list of valid human fossils in no way affects the scientific acceptance of the theory of evolution. In fact, only the circumstance that Piltdown had been discovered early, before we had a clear knowledge of the nature of human fossils and the techniques of dating them, made the long survival of this extraordinary hoax possible. Yet in the end it has been the press, absorbed in a piece of clever scientific detection, which has missed the real secret of Piltdown. Darwin saw in the rise of man, with his unique, time-spanning brain, only the undirected play of such natural forces as had created the rest of the living world of plants and animals. Wallace, by contrast, in the case of man, totally abandoned this point of view and turned instead toward a theory of a divinely directed control of the evolutionary process. The issue can be made clear only by a rapid comparison of the views of both men.

As everyone who has studied evolution knows, Darwin propounded the theory that since the reproductive powers of plants and animals potentially far outpace the available food supply, there is in nature a constant struggle for existence on the part of every living thing. Since animals vary individually, the most cleverly adapted will survive and leave offspring which will inherit, and in their turn enhance, the genetic endowment they have received from their ancestors. Because the struggle for life is incessant, this unceasing process promotes endless slow changes in bodily form, as living creatures are subjected to different natural environments, different enemies, and all the vicissitudes against which life has struggled down the ages.

Darwin, however, laid just one stricture on his theory: it could, he maintained, "render each organized being only as perfect or a little more perfect than other inhabitants of the same country." It could allow any animal only a relative superiority, never an absolute perfection—otherwise selection and the struggle for existence would cease to operate. To explain the rise of man through the slow, incremental gains of natural selection, Darwin had to assume a long struggle of man with man and tribe with tribe.

He had to make this assumption because man had far outpaced his animal associates. Since Darwin's theory of the evolutionary process is based upon the practical value of all physical and mental characters in the life struggle, to ignore the human struggle of man with man would have left no explanation as to how humanity by natural selection alone managed to attain an intellectual status so far beyond that of any of the animals with which it had begun its competition for survival.

To most of the thinkers of Darwin's day this seemed a reasonable explanation. It was a time of colonial expansion and ruthless business competition. Peoples of primitive cultures, small societies lost on the world's margins, seemed destined to be destroyed. It was thought that Victorian civilization was the apex of human achievement and that other races with different customs and ways of life must be biologically inferior to Western man. Some of them were even described as only slightly superior to apes. The Darwinians, in a time when there were no satisfactory fossils by which to demonstrate human evolution, were unconsciously minimizing the abyss which yawned between man and ape. In their anxiety to demonstrate our lowly origins they were throwing modern natives into the gap as representing living "missing links" in the chain of human ascent.

It was just at this time that Wallace lifted a voice of lonely protest. The episode is a strange one in the history of science, for Wallace had, independently of Darwin, originally arrived at the same general conclusion as to the nature of the evolutionary process. Nevertheless, only a few years after the publication of Darwin's work, *The Origin of Species*, Wallace had come to entertain a point of view which astounded and troubled Darwin. Wallace, who had had years of experience with natives of the tropical archipelagoes, abandoned the idea that they were of mentally inferior cast. He did more. He committed the Darwinian heresy of maintaining that their mental powers were far in excess of what they really needed to carry on the simple food-gathering techniques by which they survived.

"How, then," Wallace insisted, "was an organ developed so far beyond the needs of its possessor? Natural selection could only have endowed the savage with a brain a little superior to that of an ape, whereas he actually possesses one but little inferior to that of the average member of our learned societies."

At a time when many primitive peoples were erroneously assumed to speak only in grunts or to chatter like monkeys, Wallace maintained his view of the high intellectual powers of natives by insisting that "the capacity of uttering a variety of distinct articulate sounds and of applying to them an almost infinite amount of modulation . . . is not in any way inferior to that of the higher races. An instrument has been developed in advance of the needs of its possessor."

Finally, Wallace challenged the whole Darwinian position on man by insisting that artistic, mathematical, and musical abilities could not be explained on the basis of natural selection and the struggle for existence. Something else, he contended, some unknown spiritual element, must have been at work in the elaboration of the human brain. Why else would men of simple cultures possess the same basic intellectual powers which the Darwinists maintained could be elaborated only by competitive struggle?

"If you had not told me you had made these remarks," Darwin said, "I should have thought they had been added by someone else. I differ grievously from you and am very sorry for it." He did not, however, supply a valid answer to Wallace's queries. Outside of murmuring about the inherited effects of habit—a contention without scientific validity today—Darwin clung to his original position. Slowly Wallace's challenge was forgotten and a great complacency settled down upon the scientific world.

For seventy years after the publication of *The Origin of Species* in 1859, there were only two finds of fossil human skulls which seemed to throw any light upon the Darwin-Wallace controversy. One was the discovery of the small-brained Java Ape Man, the other was the famous Piltdown or "dawn man." Both were originally dated as lying at the very beginning of the Ice Age, and, though these dates were later to be modified, the skulls, for a very long time, were regarded as roughly contemporaneous and very old.

Two more unlike "missing links" could hardly be imagined. Though they were supposed to share a million-year antiquity, the one was indeed quite primitive and small-brained; the other, Piltdown, in spite of what seemed a primitive lower face, was surprisingly modern in brain. Which of these forms told the true story of human development? Was a large brain old? Had ages upon ages of slow, incremental, Darwinian increase produced it? The Piltdown skull seemed to suggest such a development.

Many were flattered to find their anthropoid ancestry seemingly removed to an increasingly remote past. If one looked at the Java Ape Man, one was forced to contemplate an ancestor, not terribly remote in time, who still had a face and a brain which hinted strongly of the ape. Yet, when by geological evidence this "erect walking ape-man" was finally assigned to a middle Ice Age antiquity, there arose the immediate possibility that Wallace could be right in his suspicion that the human brain might have had a surprisingly rapid development. By contrast, the Piltdown remains seemed to suggest a far more ancient and slow-paced evolution of man. The Piltdown hoaxer, in attaching an ape jaw to a human skull fragment, had, perhaps unwittingly, created a creature which supported the Darwinian idea of man, not too unlike the man of today, extending far back into pre-Ice Age times.

Which story was the right one? Until the exposé of Piltdown in 1953, both theories had to be considered possible and the two hopelessly unlike fossils had to be solemnly weighed in the same balance. Today Piltdown is gone. In its place we are confronted with the blunt statement of two modern scientists, M. R. A. Chance and A. P. Mead.

"No adequate explanation," they confess over eighty years after Darwin scrawled his vigorous "No!" upon Wallace's paper, "has been put forward to account for so large a cerebrum as that found in man." [1]

We have been so busy tracing the tangible aspects of evolution in the *forms of animals* that our heads, the

[1] *Symposia of the Society for Experimental Biology*, VII, Evolution (New York: Academic Press, 1953), p. 395.

little globes which hold the midnight sky and the shining, invisible universes of thought, have been taken about as much for granted as the growth of a yellow pumpkin in the fall.

Now a part of this mystery as it is seen by the anthropologists of today lies in the relation of the brain to time. "If," Wallace had said, "researches in all parts of Europe and Asia fail to bring to light any proofs of man's presence far back in the Age of Mammals, *it will be at least a presumption that he came into existence at a much later date and by a more rapid process of development.*" If human evolution should prove to be comparatively rapid, "explosive" in other words, Wallace felt that his position would be vindicated, because such a rapid development of the brain would, he thought, imply a divinely directed force at work in man. In the 1870's when he wrote, however, human prehistory was largely an unknown blank. Today we can make a partial answer to Wallace's question. Since the exposure of the Piltdown hoax all of the evidence at our command—and it is considerable—points to man, in his present form, as being one of the youngest and newest of all earth's swarming inhabitants.

The Ice Age extends behind us in time for, at most, a million years. Though this may seem long to one who confines his studies to the written history of man, it is, in reality, a very short period as the student of evolution measures time. It is a period marked more by the extinction of some of the last huge land animals, like the hairy mammoth and the saber-toothed tiger, than it is by the appearance of new forms of life. To this there is only one apparent exception: the rise and spread of man over the Old World land mass.

Most of our knowledge of him—even in his massive-faced, beetle-browed stage—is now confined, since the loss of Piltdown, to the last half of the Ice Age. If we pass backward beyond this point we can find traces of crude tools, stone implements which hint that some earlier form of man was present here and there in Europe, Asia, and particularly Africa in the earlier half of Ice Age time, but to the scientist it is like peering into the mists floating over an unknown landscape. Here and there through the swirling vapor one catches a glimpse of a shambling figure, or a half-wild primordial face stares back at one from some momentary opening in the fog. Then, just as one grasps at a clue, the long gray twilight settles in and the wraiths and the half-heard voices pass away.

Nevertheless, particularly in Africa, a remarkable group of human-like apes have been discovered: creatures with small brains and teeth of a remarkably human cast. Prominent scientists are still debating whether they are on the direct line of ascent to man or are merely near relatives of ours. Some, it is now obvious, existed too late in time to be our true ancestors, though this does not mean that their bodily characters may not tell us what the earliest anthropoids who took the human turn of the road were like.

These apes are not all similar in type or appearance. They are men and yet not men. Some are frailer-bodied, some have great, bone-cracking jaws and massive gorilloid crests atop their skulls. This fact leads us to another of Wallace's remarkable perceptions of long ago. With the rise of the truly human brain, Wallace saw that man had transferred to his machines and tools many of the alterations of parts that in animals take place through evolution of the body. Unwittingly, man had assigned to his machines the selective evolution which in the animal changes the nature of its bodily structure through the ages. Man of today, the atomic manipulator, the aeronaut who flies faster than sound, has precisely the same brain and body as his ancestors of twenty thousand years ago who painted the last Ice Age mammoths on the walls of caves in France.

To put it another way, it is man's ideas that have evolved and changed the world about him. Now, confronted by the lethal radiations of open space and the fantastic speeds of his machines, he has to invent new electronic controls that operate faster than his nerves, and he must shield his naked body against atomic radiation by the use of protective metals. Already he is physically antique in this robot world he has created. All that sustains him is that small globe of gray matter through which spin his ever-changing conceptions of the universe.

Yet, as Wallace, almost a hundred years ago, glimpsed this timeless element in man, he uttered one more prophecy. When we come to trace out history into the past, he contended, sooner or later we will come to a time when the body of man begins to differ and diverge more extravagantly in its appearance. Then, he wrote, we shall know that we stand close to the starting point of the human family. In the twilight before the dawn of the human mind, man will not have been able to protect his body from change and his remains will bear the marks of all the forces that play upon the rest of life. He will be different in his form. He will be, in other words, as variable in body as we know the South African man-apes to be.

Today, with the solution of the Piltdown enigma, we must settle the question of the time involved in human evolution in favor of Wallace, not Darwin; we need not, however, pursue the mystical aspects of Wallace's thought—since other factors yet to be examined may well account for the rise of man. The rapid fading out of archaeological evidence of tools in lower Ice Age times—along with the discovery of man-apes of human aspect but with ape-sized brains, yet possessing a diverse array of bodily characters—suggests that the evolution of the human brain was far more rapid than that conceived of in early Darwinian circles. At that time it was possible to hear the Eskimos spoken of as possible survivals of Miocene men of several million years ago. By contrast to this point of view, man and his rise now appear short in time—explosively short. There is every reason to believe that whatever the nature of the forces involved in the production of the human brain, a long slow competition of human group with human group or race with race would not have resulted in such similar mental potentialities among all peoples everywhere. Something—some other factor—has escaped our scientific attention.

There are certain strange bodily characters which mark man as being more than the product of a dog-eat-dog competition with his fellows. He possesses a peculiar larval nakedness, difficult to explain on survival principles; his periods of helpless infancy and childhood are prolonged; he has aesthetic impulses which, though they vary in intensity from individual to individual, appear in varying manifestations among all peoples. He is totally dependent, in the achievement of human status, upon the careful training he receives in human society.

Unlike a solitary species of animal, he cannot develop alone. He has suffered a major loss of precise instinctive controls of behavior. To make up for this biological lack, society and parents condition the infant, supply his motivations, and promote his long-drawn training at the difficult task of becoming a normal human being. Even today some individuals fail to make this adjustment and have to be excluded from society.

We are now in a position to see the wonder and terror of the human predicament: man is totally dependent on society. Creature of dream, he has created an invisible world of ideas, beliefs, habits, and customs which buttress him about and replace for him the precise instincts of the lower creatures. In this invisible universe he takes refuge, but just as instinct may fail an animal under some shift of environmental conditions, so man's cultural beliefs may prove inadequate to meet a new situation, or, on an individual level, the confused mind may substitute, by some terrible alchemy, cruelty for love.

The profound shock of the leap from animal to human status is echoing still in the depths of our subconscious minds. It is a transition which would seem to have demanded considerable rapidity of adjustment in order for human beings to have survived, and it also involved the growth of prolonged bonds of affection in the subhuman family, because otherwise its naked, helpless offspring would have perished.

It is not beyond the range of possibility that this strange reduction of instincts in man in some manner forced a precipitous brain growth as a compensation—something that had to be hurried for survival purposes. Man's competition, it would thus appear, may have been much less with his own kind than with the dire necessity of building about him a world of ideas to replace his lost animal environment. As we will show later, he is a pedomorph, a creature with an extended childhood.

Modern science would go on to add that many of the characters of man, such as his lack of fur, thin skull, and globular head, suggest mysterious changes in growth rates which preserve, far into human maturity, foetal or infantile characters which hint that the forces creating man drew him fantastically out of the very childhood of his brutal forerunners. Once more the words of Wallace come back to haunt us: "We may safely infer that the savage possesses a brain capable, if cultivated and developed, of performing work of a kind and degree far beyond what he ever requires it to do."

As a modern man, I have sat in concert halls and watched huge audiences floating dazed on the voice of a great singer. Alone in the dark box I have heard far off as if ascending out of some black stairwell the guttural whisperings and bestial coughings out of which that voice arose. Again, I have sat under the slit dome of a mountain observatory and marveled, as the great wheel of the galaxy turned in all its midnight splendor, that the mind in the course of three centuries has been capable of drawing into its strange, nonspatial interior that world of infinite distance and multitudinous dimensions.

Ironically enough, science, which can show us the flints and the broken skulls of our dead fathers, has yet to explain how we have come so far so fast, nor has it any completely satisfactory answer to the question asked by Wallace long ago. Those who would revile us by pointing to an ape at the foot of our family tree grasp little of the awe with which the modern scientist now puzzles over man's lonely and supreme ascent. As one great student of paleoneurology, Dr. Tilly Edinger, recently remarked, "If man has passed through a Pithecanthropus phase, the evolution of his brain has been unique, not only in its result but also in its tempo. . . . Enlargement of the cerebral hemispheres by 50 per cent seems to have taken place, speaking geologically, within an instant, and without having been accompanied by any major increase in body size."

The true secret of Piltdown, though thought by the public to be merely the revelation of an unscrupulous forgery, lies in the fact that it has forced science to reëxamine carefully the history of the most remarkable creation in the world—the human brain.

THE MAZE

✟

Shortly after I had expressed my conclusions about the real secret of Piltdown, I was roundly castigated by a few people who had construed my remarks as an attack upon Darwin and thus an assault upon the theory of evolution itself. A surprising amount of suppressed emotion still lingers about these hundred-year-old controversies, and those who are not historical-minded may be quick to launch themselves, sometimes more valiantly than accurately, into the thick of some forgotten fray. Along an advancing front of science, the man who writes for a nontechnical public runs risks and has curious experiences. Sometimes he is unlucky, as in the case of an acquaintance of mine whose article dealing soberly with the Piltdown skull appeared at the very moment when the hoax was denounced in the press. Sometimes, on the other hand, he may have almost preternatural luck, in that unexpected events may further substantiate a view that he had earlier broached in hesitation and with a minimum of supporting evidence.

After I had expressed myself upon the dangerously controversial subject of the human brain—and I say this avowedly, though so distinguished an authority upon the great apes as Solly Zuckerman has spoken of the "enormous gap" which exists "between the intelligence of Man and that of any other Primate"—two quite astonishing things happened. The first of these it is my intention to chronicle in this chapter; the second event, and the final culmination of the plot, will have to be reserved for the one which follows. The first happening, as it was described in the press, seemed to be a total negation of much that has been expressed in my treatment of the Piltdown story, namely, the recency of man.

The reader may remember that in March of 1956 curious and startling headlines began to appear in the newspapers. In the first excitement it must have seemed to the layman that the whole theory of evolution was about to be overthrown. There were accounts in the press of a ten-million-year-old "human" fossil. Such a discovery seemed, at first thought, to contradict what I had contended was the great youth of man, that is, man as a culture bearer, a user of speech.

The commotion had been touched off by the arrival in New York City of a paleontologist from Switzerland bearing the bones of a small primate long known to science as Oreopithecus. Johannes Hurzeler of Basel presented to a group of scholars gathered at the Wenner-Gren Foundation for Anthropological Research his view that the bones of Oreopithecus showed human rather than anthropoid affinities. Since these bones are estimated to be ten million years older than the earliest known fossil men, his announcement made headlines.

"Fossil Research Questions Darwin Evolution Theory," the *New York Times* announced. The *Herald Tribune* editorialized: "No Missing Link?" Specialists on fossil man were besieged by telephone calls from reporters and by faintly derisive queries from anti-evolutionists whose interest had already been whetted by the Piltdown hoax. Perhaps this new contradiction would mark the final exit of the man-monkey and of the anthropologists along with it.

By the time scientists had begun to respond, the press had passed on to other things, leaving in the mind of the public a confused vision of a sort of "little man" who, so the newspapers said, had been found in a coal mine in Tuscany. Like most such episodes, that of Oreopithecus has a history, and the argument over it is of the same general nature as two similar controversies fought within the memory of men now living.

The incident has served to draw attention to a long-existing debate among anthropologists, which has occasionally waxed acrimonious. The partisans divide basically into two schools: the school of the "little man" and that of the "apeman." The former pursue the figure of man backward until, upon some far wall in time, it appears as a dwarfed, big-headed little shadow; the latter see our earliest ancestor shambling into the light like some great shaggy anthropoid. The argument recalls the ancient dispute between the preformationists, who saw in the human sperm cell a preformed homunculus, or little man, which had only to grow to adult size, and the epigenesists, who judged correctly that each embryo acquires the characteristics of a human being only through development.

Some anthropologists search for human characters—vertical front teeth, a shortened face, an expanded brain case—early in the human line of descent. They seek, in other words, for something dangerously close to the homunculus of the preformationists. They "prove" evolution by finding, as St. George Jackson Mivart said in 1874, "an ancestral form so like man [that] we have the virtual pre-existence of man's body supposed, in order to account for the actual first appearance of that body as we know it."

The more thoroughgoing evolutionists, in contrast,

have looked for forms which contained only the *possibility* of development into man. Such students have generally regarded man as a relatively recent emergent from a group of primates which also gave rise to the modern great apes; in other words, the comparison of man with the anthropoids of today has been based on the assumption that they and we had ancestors in common.

Charles Darwin was not the first to notice our likeness to the monkeys and apes. Such observations extend into antiquity, and by the eighteenth and early nineteenth centuries philosophers were arranging the primates in an order of complexity. As voyagers began to come into contact with primitive peoples, these were often placed on the scale as grades between the anthropoids and civilized European man. The Hottentots of the Cape of Good Hope particularly appealed to the Western mind as candidates for such a place; it was said that their language was only a step above the chatter of apes.

Thus notions of the "missing link" were in existence long before Darwin and long before the appearance of a truly evolutionary philosophy. Darwin himself cautiously refrained from attempting to trace man's precise relationship to the apes. But some of his followers, notably T. H. Huxley, tackled the problem head on. Huxley was provoked to his excursion into man's past by events at the famous meeting of the British Association for the Advancement of Science at Oxford in 1860. He had borne the brunt of the conservatives' attacks on evolution. At this meeting Richard Owen, England's foremost comparative anatomist and a mortal enemy of Darwin and his followers, attempted to maintain man's unique position in the animal world by placing him in a distinct subclass of the mammals for which he proposed the name "Archencephala." This classification was based upon brain characters which Owen maintained did not occur in the lower primates. Huxley, his ire aroused, set out to demonstrate that Owen was wrong, that man was closely related to the other primates. He composed a series of lectures which were published in 1863 under the title *Evidence as to Man's Place in Nature.*

In this work, which more or less set the pattern for much that followed, Huxley thoroughly demolished Owen's position. He took the view that "the surface of the brain of a monkey exhibits a sort of skeleton map of man's, and in the manlike apes the details become more and more filled in, until it is only in minor characters . . . that the chimpanzee's or the orang's brain can be structurally distinguished from man's." Huxley was quite willing to admit that man's own origin was obscure and might go back millions of years to a common ancestor, but he insisted that the modern apes were our closest surviving relatives. If Huxley dwelt too heavily and too emotionally upon anatomical correspondence between ourselves and the great apes, it must be remembered that at the time he wrote the evolutionists were fighting primarily for a principle, against the orthodox "special creationists." Furthermore, it must also be remembered that very few human fossils had been discovered, and these were fragmentary. Our living relatives in the trees could be seen at the zoo, and it was inevitable that they should dominate man's imagination. Serious scholars even came to believe that microcephalic idiots were throwbacks to some remote period of the human past.

By the beginning of the twentieth century the ape origins of modern man seemed pretty well established. The finding of the Pithecanthropus skull cap had bolstered this view. Many felt that from a form something like that of a chimpanzee it was an easy step to the Java man and thence on to Neanderthal and modern man. But at the turn of the century there came a new revolt against the ape.

The attention of anatomists was attracted to a small, tree-living creature in southeast Asia possessing definite characters of a primate. The tarsier (*Tarsius spectrum*), an animal with enormous eyes and about the size of a small kitten, has a brain and other characteristics which ally it to the lower monkeys. In 1918, F. Wood Jones, a distinguished English anatomist, had expressed the heretical view, which he has maintained and developed since, that man arose from a tarsioid rather than from an anthropoid ancestry.

Wood Jones insists that the human line is very ancient, going back to a past tens of millions of years old in the Tertiary Period. He predicts that man's immediate ancestors, if ever discovered, "will be utterly unlike the slouching, hairy 'ape-men' of which some have dreamed . . . and will be found in geological strata antedating the heyday of the great apes." The ancestors of man, he says, were "small, active animals" already endowed with legs longer than their arms, small jaws without protruding teeth, and enlarged craniums. They were not swingers in trees: the human hand and foot, he contends, are too specialized to have been made over rapidly from an arboreal ancestor's. The present-day tarsiers in the trees, according to his view, evolved their tree-living specializations later, but our early tarsioid ancestor walked on the ground.

Wood Jones's proto-man thus sounds like a homunculus. When he first advocated his views, he found very few followers. Henry Fairfield Osborn, the late paleontologist, though not a Wood Jones follower, inclined toward a homuncular dawn man going back to early Tertiary times many millions of years ago. "I predict," he said, "that even in Upper Oligocene time we shall find pro-men, and that they will have pro-human limbs."

Wood Jones and Osborn were vigorously refuted by primatologists who championed the orthodox view that man was a "made-over ape." They insisted that man's immediate forerunners could not be so ancient as Wood Jones and Osborn said. "It seems anachronistic," wrote William King Gregory, "to attribute to the very remote Tertiary ancestors of man the long legs, long thumbs, big brain, short face, small canines, etc., which are now diagnostic characters." But by the 1940's the "made-over ape" point of view had moderated. The most important factor in this change was the discovery in South Africa of the fossil *Proconsul africanus*—a creature of the early Miocene (about twenty million years ago) which combined characters of early Old World monkeys and great apes. William L. Straus, Jr., of the Johns Hopkins University, voiced a suspicion that man's immediate ancestors might have been "more monkey-like than anthropoid-like." Straus, who takes a very sane and cautious position on this lengthy controversy over the human ancestry, feels that the anthropoid-ape theory is weakest in its failure to account for anatomical traits which man shares with the monkeys and lemurs. More recently W. C. Osman Hill, the well-known English primatologist, has come to believe that man branched off the primate stock below the great-ape line. He even suggests that Straus's view might be reconciled with Wood Jones's tarsioid hypothesis if some early Oligocene monkey of tarsioid affinities were admitted on the line leading to man—a form, say, like Parapithecus.

Thus, before Hurzeler's recent announcement a slow

shift of thought or widening of possible horizons had been under way in the study of human evolution. The theory that man came down late out of the trees has been dropped in some quarters and is less explosively defended in others. There is a greater willingness to reserve judgment and wait upon new evidence. It was in this receptive atmosphere that Hurzeler presented his new study of Oreopithecus.

The fossil has been known since 1872, when it was described by the French paleontologist Paul Gervais, who regarded it as an Old World monkey. Hurzeler, after studying the original fossil and later finds, has become convinced that Oreopithecus is the first manlike form discovered in the Tertiary Period—it is believed to date from the Miocene. He apparently bases this view upon certain technical features of the teeth, including the nonprojecting canines, the vertical bite and the shortened face. It must be noted, however, that only parts of the skull have been found, and its full shape cannot be reconstructed.

Oreopithecus is a lower "monkey," in popular terms. It is not a "man" in the sense that many reporters assumed it to be, in spite of "no tooth gaps, no apelike protruding jaw," and so on. There are both fossil and still living primates which would have no trouble in answering that description, yet I am sure no one would call them men.

So the substance of the story is that Hurzeler has revived interest in a problematical bit of bone we have long been fingering. For the successful reconstruction of the evolution of the horse in the Tertiary Period, paleontologists had thousands of fossil bones to study. Primatologists may therefore be forgiven their fumblings over great gaps of millions of years from which we do not possess a single complete monkey skeleton, let alone the skeleton of a human forerunner. For the whole Tertiary Period, which involves something like sixty to eighty million years, we have to read the story of primate evolution from a few handfuls of broken bones and teeth. Those fossils, moreover, are from places thousands of miles apart on the Old World land mass.

If we were able to follow every step of man's history backward into time, we would see him divested, rag by rag and stitch by stitch, of every vestige of his human garment. That divestment, however, would not occur all at one place. If we accept the evidence of evolution, we must assume that man became man by degrees, that he emerged out of the animal world by the slow accumulation of human characters over long ages—save for that seemingly rapid spurt in brain growth, which has carried him so far from his other relatives.

Our knowledge at present is not sufficient to establish precisely what anatomical traits are peculiarly human. As Straus has very aptly pointed out: "It is this general lack of structural specialization that makes the study of primate phylogeny so difficult." Some traits may have been paralleled in primate lines of evolution which did not lead to man; some traits called human may represent old generalized characters which have survived in man and been lost in some of his modern specialized relatives.

To continue our writing of the story of human evolution we are totally dependent upon finding additional fossils. Until further discoveries accumulate, each student will perhaps inevitably read a little of his own temperament into the record. Some, as Hurzeler has done, will dwell upon short faces, vertical front teeth and little rounded chins. They will catch glimpses of an elfin human figure which mocks us from a remote glade in the forest of time. Others, just as competent, will say that this elusive homuncular elf is a dream spun from our disguised human longing for an ancestor like ourselves. They will say that in the living primate world around us there are lemurs with short faces and vertical teeth, and that there are monkeys which have the genuine faces of elves and the capacious craniums of little men.

In the end we may shake our heads, baffled, and have to admit that many lines of seeming relatives, rather than merely one, lead to man. It is as though we stood at the heart of a maze and no longer remembered how we had come there.

THE DREAM ANIMAL

🌳

It will now be seen that in spite of the dramatic press announcements which thundered the end of Darwinism and of missing links, our little homuncular elf proved nothing of the kind. Even if he turned out to be on the main line of evolutionary ascent leading to man—and this is still exceedingly doubtful—he has, at present, nothing to tell us about the human brain. He is small, he is not by any stretch of imagination a man, and if he did indeed become one, the event still lay millions of years in the future. No amount of headlines can turn the little creature from Tuscany into a human being without recourse to evolutionary change. The writers who had seized upon the "little man" as a refutation of Darwin's general thesis had, at best, been merely acclaiming a new "missing link."

We must now examine, however, some recent aspects of the problem to which I have previously given attention: the mystery which enshrouds the rise of the human brain. A most perceptive philosopher once remarked that the truth about man is inside him. This may well prove to be the case, but the difficulty is to get the secret out, if indeed it lies there, and once it is revealed, to be sure that it is read correctly.

Every so often out of the millions of the human population, a six-year-old child or a teen-age youth dies of old age. The cause of this curious disease, known as progeria, or premature aging, is totally unknown. Clinical cases are reported of complete hairlessness, wrinkled and flabby skin, along with senile changes in the heart and blood vessels. Medical science has observed in these rare cases an enormous increase in the velocity of aging, but the mechanism involved remains as yet undiscovered, though the cause may lie somewhere among the ductless glands.

The affliction, rare though it is, reveals a mysterious clock in the body, a clock capable of running fast or slow, shortening life or extending it and, like the more visible portions of our anatomy, being subjected to evolutionary selection. This clock, however, has another even more curious aspect: it may affect the growth rate of particular organs. In this way certain peculiar animal specializations have appeared, such as the huge antlers of the extinct Irish elk, or the dagger-like fangs of the saber-toothed tiger.

Man, too, has a curious specialization of a more abstract and generalized type, his brain. If this brain, a brain more than twice as large as that of a much bigger animal—the gorilla—is to be acquired in infancy, its major growth must take place with far greater rapidity than in the case of man's nearest living relatives, the great apes. It must literally spring up like an overnight mushroom, and this greatly accelerated growth must take place during the first months after birth. If it took place in the embryo, man would long since have disappeared from the planet—it would have been literally impossible for him to have been born. As it is, the head of the infant is one of the factors making human birth comparatively difficult. When we are born, however, our brain size, about 330 cubic centimeters, is only slightly larger than that of a gorilla baby. This is why human and anthropoid young look so appealingly similar in their earliest infancy.

A little later, an amazing development takes place in the human offspring. In the first year of life its brain trebles in size. It is this peculiar leap, unlike anything else we know in the animal world, which gives to man his uniquely human qualities. When the leap fails, as in those rare instances where the brain does not grow, microcephaly, "pinheadedness," is the result, and the child is then an idiot. Somewhere among the inner secrets of the body is one which keeps the time for human brain growth. If we compare our brains with those of other primate relatives (recognizing, as we do, many similarities of structure) we are yet unable to perceive at what point in time or under what evolutionary conditions the actual human forerunner began to manifest this strange postnatal brain expansion. It has carried him far beyond the mental span of his surviving relatives. As our previously quoted authority, Dr. Tilly Edinger of Harvard, has declared, "the brain of *Homo Sapiens* has not evolved from the brains it is compared with by comparative anatomy; it developed within the Hominidae, at a late stage of the evolution of this family whose other species are all extinct."

We can, in other words, weigh, measure and dissect the brains of any number of existing monkeys. We may learn much in the process, but the key to our human brain clock is not among them. It arose in the germ plasm of the human group alone and we are the last living representatives of that family. As we contemplate, however, the old biological law that, to a certain degree, the history of the development of the individual tends to reproduce the evolutionary history of the group to which it belongs, we cannot help but wonder if this remarkable spurt in brain development may not represent something roughly akin to what happened in the geological past of man—a sudden or explosive increase which was achieved in a relatively short period, geologically speaking. We have already opened this topic in our discussion of the Darwin-Wallace argument. Let us now see what new evidence bears upon the facts we set forth there.

In discussing the significance of the Piltdown hoax and its bearing upon the Darwin-Wallace controversy, I used the accepted orthodox geological estimate of the time involved in that series of fluctuating events which we speak of popularly as the "Ice Age." I pointed out that almost all of what we know about human evolution is confined to this period. Long though one million years may seem compared with our few millennia of written history, it is, in geological terms, in evolutionary terms, a mere minute's tick of the astronomical clock.

Among other forms of life than man, few marked transformations occurred. Rather, the Ice Age was, particularly toward its close, a time of great extinctions. Some of the huge beasts whose intercontinental migrations had laid down the first paths along which man had traveled, vanished totally from the earth. Mammoths, the Temperate Zone elephants, dropped the last of their heavy tusks along the receding fringes of the ice. The long-horned bisons upon whose herds man had nourished himself for many a long century of illiterate wanderings, faded back into the past. The ape whose cultural remnants at the beginning of the first glaciation can scarcely be distinguished from chance bits of stone has, by the ending of the fourth ice, become artist and world rover, penetrator of the five continents, and master of all.

There is nothing quite like this event in all the time that went before; the end of brute animal dominance upon earth had come at last. For good or ill, the growth of forests or their destruction, the spread of deserts or their elimination, would lie more and more at the whim of that cunning and insatiable creature who slipped so mysteriously out of the green twilight of nature's laboratory a short million years ago.

A million years is a short time as evolution clocks its progress. We assume, of course, that below that point the creature which was to become man was still walking on his hind feet, but there is every reason to think that the bulging cortex which would later measure stars and ice ages was still a dim, impoverished region in a skull box whose capacity was no greater than that of other apes. Still, a million years in the life history of a single active species like man is a long time, and powerful selective forces must have been at work as ice sheets ground their way across vast areas of the temperate zones. But suppose, just suppose for a moment, that this period of the great ice advances did not last a million years—suppose our geological estimates are mistaken. Suppose that this period we have been estimating at one million years should instead have lasted, say, a third of that time.

In that case, what are we to think of the story of man? Into what foreshortened and cramped circumstances is the human drama to be reduced, a drama, moreover, which, besides evolutionary change, involves time for the spread of man into the New World? Such an episode, it is obvious, would involve a complete reëxamination of our thinking upon the subject of human evolution. In 1956 Dr. Cesare Emiliani of the University of Chicago introduced just this startling factor into the dating of the Ice Age. He did it by the application of a new dating process developed in the field of atomic physics.[1]

The method, it should be explained at the outset, is not the carbon-14 technique which has become so widely publicized in the last decade. That method has applications which, at best, can carry us back around thirty to forty thousand years. The new technique elaborated in the University of Chicago laboratories involves oxygen-18. By studying the amount of this isotope in the shells of sea creatures it was found that the percentage of oxygen-18 in the limy shell of, say, an oyster would reveal the temperature of the water in which the oyster had lived when its shell was being secreted. This

[1] "Note on Absolute Chronology of Human Evolution," *Science* 123 (1956), pp. 924-26.

is because oxygen-18 enters chemical reactions differently at different temperatures. For example, as the temperature of the water increases, the oxygen-18 in the shell decreases.

By using marine cores, specimens of undisturbed sediments brought up from the ocean floor, Dr. Emiliani has been able to subject these chalky deposits full of tiny shells to careful oxygen-18 analysis. He has found, as he analyzed the chemical nature of the seas' "long snowfall," that is, the age-long rain of microscopic shells falling gently to the sea bottom, that marked changes in water temperature could be discerned for different periods in the past. As he studied layer after layer of the chalky ooze brought up in sequential order from the depths, he found that the times of maximum ice expansion on the continents coincided with periods of marked cold beyond that of the present, as revealed in the oxygen-18 content of the minute shells from the ocean floor.

Studying Atlantic and Caribbean cores, Emiliani came to the conclusion that the earliest great cold period, most probably coinciding with the onset of the first glaciation in Europe, was probably no earlier than about three hundred thousand years ago. Oxygen-18, of course, indicates periods of relative warmth or cold, not years. The dating triumph was achieved by the well-known carbon-14 technique for the upper levels of the deposits within the forty-thousand-year range, since carbon-14 also occurs in the chalk ooze.

By establishing the beginning of the last ice recession at about twenty thousand years, it was possible, as a result of the undisturbed uniform nature of the sea deposits, to project the datings backward by the combination of the cold graph and the apparent rate at which the deposits had been laid down, as determined from the carbon dates of the more recent levels. The study reveals a considerable degree of regularity in the waxing and waning of the ice sheets at intervals of about fifty to sixty thousand years.

Dr. Emiliani and his co-workers have thus produced an Ice Age chronology startlingly different from orthodox estimates, but one which is being widely and favorably considered. The newer scheme allows about six hundred thousand years for the total of Ice Age time. Actually the modification is more striking than this figure would indicate. Older figures placed the first, or Gunz glaciation, distant from us at the bottom of the Ice Age by almost a million years. The new chronology would place this ice sheet only about three hundred thousand years remote and then allow perhaps three hundred thousand more years, much less accurately computable and quite indefinite, for certain vague preglacial events. These might include our oldest traces of the Australopithecine man-apes and the first dim traces of crude pebble and bone tools, possibly made by some, at least, of these South African anthropoids.

As we have already indicated, most of our collection of human fossils is derived from the last half of the Pleistocene, even by the old chronology. In this new arrangement the bulk of this material is found to be less than two hundred thousand years old. Man, in Dr. Emiliani's own words, had "the apparent ability to evolve rapidly." This is almost an understatement. The new chronology would appear to suggest a spectacular, even more explosive development than I have previously suggested.

Unfortunately the full outlines of this story cannot, as yet, be made out. Our fossils are too scattered and too few. If the Fontechevade cranium from the French third interglacial represents a man essentially like ourselves, as in brain he appears to be, we can date our species as in existence perhaps seventy thousand years ago, though its total diffusion, in terms of area, at that date would be unknown. If the problematical Swanscombe skull—discovered in England—whose face is missing but whose cranial capacity falls within the modern range, should prove, in time, to be also of our own species, "modern" man would have been in existence perhaps one hundred twenty thousand years before the present.

Even if the men of this period should, in the end, prove to have a face somewhat more massive than that of modern man, an essentially modern brain at so early a date can only suggest, in the light of Emiliani's new datings, that the rise of man from a brain level represented in earliest preglacial times by the South African man-apes took place with extreme rapidity. Either this occurred, or other fossil forms are not on the main line of human ascent at all. This latter theory, if we still try to cling to a slow type of human evolution, would imply that the true origin of our species is lost in some older pre-Ice Age level, and that all the other human fossils represent side lines and blind alleys of development, living fossils already archaic in Pleistocene times.

Some, contending for this view, have pointed out that carbon-14 datings close to the forty-thousand-year mark have recently been recorded in America. This, it has been argued, suggests a remarkably wide and early diffusion for man, if he is really so young as is now suggested. Just lately, however, some of the earliest carbon-14 dates from the Southwest have been challenged. Professor Frederick Zeuner of the University of London has recently (1957) reported that carbon samples subjected to alkaline washing give dates much earlier than they should actually be. Some of the carbon-14 necessary for accurate dating is apparently removed by subjection to this treatment, thus raising the age of the sample. As a consequence, some of the very earliest American dates from the Southwest may be subject to upward revision. There is no doubt that man had reached America in the closing Ice Age, but these earlier dates will be subject to serious scrutiny.

Interestingly enough, the Keilor skull from Australia, once supposed to be a very early third interglacial man of our own species, has now been elevated, on the basis of carbon dates, to definitely postglacial times. Thus, on this remote continent, there is now no reliable evidence of extremely ancient human intrusion. Furthermore, if we turn to the Old World and seek to carry men much like ourselves further back toward the first glaciation, we have to ask why we so rapidly descend into seemingly cultureless or almost cultureless levels. If man approximating ourselves is truly much older than we imagine, it is conceivable that his physical remains might for long escape us. It seems unlikely, however, that a large-brained form, if widely diffused, would have left so little evidence of his activities. It would appear, then, that within the very brief period between about five hundred thousand to one hundred fifty thousand years ago, man acquired the essential features of a modern brain. Admittedly the outlines of this process are dim, but all the evidence at our command points to this process as being surprisingly rapid.

Such rapidity suggests other modes of selection and evolution than those implied in the nineteenth-century literature with its emphasis on intergroup "struggle" which, in turn, would have demanded large populations. We must make plain here, however, that to reject the older Darwinian arguments is not necessarily to reject the principle of natural selection. We may be simply dealing with a situation in which both Darwin and Wallace failed, in different ways, to see what selective forces might be at work in man. Most of the Victorian biologists were heavily concerned with the more visible aspects of the struggle for existence. They saw it in the ruthless, expanding industrialism around them; they tended to see nature as totally "red in tooth and claw."

The anthropologist had yet to subject native societies to careful scrutiny, or to learn that people of different cultures were remarkably like ourselves in their basic mental make-up. They were often regarded as mentally inferior, living fossils pushed to the wall and going under in the struggle with the dominating white. Wallace, as we have already seen, stood somewhat outside this Victorian prejudice, and having himself endured economic want, almost alone among the great biologists of his time, sought for another key to the development of man.

His thoughts led him in a somewhat mystical direction, yet certain of the facts he recorded were valid enough. He wrote early, however, so that natural explanations which could now be offered were, understandably, not available to him at that time. It is impressive that Wallace observed, though he did not understand, what we today call the pedomorphic features of man—his almost hairless body, his helpless childhood, his surprisingly developed brain—which he rightly judged to be in some manner related to the uniqueness of man. His conclusion that the linguistic ability of natives is in no way inferior to that of "higher" races—a commonplace today—was, in its own time, a courageous statement made in considerable contradiction to beliefs widely held even among scientists.

Although there is still much that we do not understand, it is likely that the selective forces working upon the humanization of man lay essentially in the nature of the socio-cultural world itself. Man, in other words, once he had "crossed over" into this new invisible environment, was being as rigorously selected for survival within it as the first fish that waddled up the shore on its fins. I have said that this new world was "invisible." I do so advisedly. It lay, not so much in his surroundings as in man's brain, in his way of looking at the world around him and at the social environment he was beginning to create in his tiny human groupings.

He was becoming something the world had never seen before—a dream animal—living at least partially within a secret universe of his own creation and sharing that secret universe in his head with other, similar heads. Symbolic communication had begun. Man had escaped out of the eternal present of the animal world into a knowledge of past and future. The unseen gods, the powers behind the world of phenomenal appearance, began to stalk through his dreams.

Nature, one might say, through the powers of this mind, grossly superstitious though it might be in its naïve examination of wind and water, was beginning to reach out into the dark behind itself. Nature was beginning to evade its own limitations in the shape of this strange, dreaming and observant brain. It was a weird multiheaded universe, going on, unseen and immaterial save as its thoughts smoldered in the eyes of hunters huddled by night fires, or were translated into pictures upon cave walls, or were expressed in the trappings of myth or ritual. The Eden of the eternal present that the animal world had known for ages was shattered at last. Through the human mind, time and darkness, good and evil, would enter and possess the world.

The Victorian biologists, intent upon the nature of the animal struggle for existence, in some degree misread human society and the kind of social selection toward brain enhancement which would be the product of unceasing struggle, not by ax and spear in the war of nature, but in that world of streaming shadows forever hidden behind the forehead of man. It was a struggle for symbolic communication, for in this new societal world communication meant life. The world of instinct was passing. This emergent creature was not whole, was not made truly human until, in infancy, the dreams of the group, the social constellation amidst which his own orbit was cast, had been implanted in the waiting, receptive substance of his brain.

How did this brain first come? How fast did it come? Probing among rocks and battered skulls, scientists find that the answers are few. There are many living members of the primate order—that order which includes man—who live in groups, but show no signs of becoming men. Their brains bear a family resemblance to our own, but they are not the brains of men. They contain, instead, only the shrewd, wild thoughts that serve to remind us of the solitary door which began to open for us once, and once only, long ago, as the earth swung in some tilted, sunlit orbit far backward on the roads of space.

If one attempts to read the complexities of the story, one is not surprised that man is alone on the planet. Rather, one is amazed and humbled that man was achieved at all. For four things had to happen, and if they had not happened simultaneously, or at least kept pace with each other, the bones of man would lie abortive and forgotten in the sandstones of the past:

1. His brain had almost to treble in size.

2. This had to be effected, not in the womb, but rapidly, after birth.

3. Childhood had to be lengthened to allow this brain, divested of most of its precise instinctive responses, to receive, store, and learn to utilize what it received from others.

4. The family bonds had to survive seasonal mating

and become permanent, if this odd new creature was to be prepared for his adult role.

Each one of these major points demanded a multitude of minor biological adjustments, yet all of this—change of growth rate, lengthened age, increased blood supply to the head, moved apparently with rapidity. It is a dizzying spectacle with which we have nothing to compare. The event is complex, it is many-sided, and what touched it off is hidden under the leaf mold of forgotten centuries.

Somewhere in the glacial mists that shroud the past, Nature found a way of speeding the proliferation of brain cells and did it by the ruthless elimination of everything not needed to that end. We lost our hairy covering, our jaws and teeth were reduced in size, our sex life was postponed, our infancy became among the most helpless of any of the animals because everything had to wait upon the development of that fast-growing mushroom which had sprung up in our heads.

Now in man, above all creatures, brain is the really important specialization. As Gavin de Beer, Director of the British Museum of Natural History, has suggested, it appears that if infancy is lengthened, there is a correspondingly lengthier retention of embryonic tissues capable of undergoing change.[2] Here, apparently, is a

---

[2] *Embryos and Ancestors,* rev. ed. (New York, Oxford, 1951), p. 93.

possible means of stepping up brain growth. The anthropoid ape, because of its shorter life cycle and slow brain growth, does not make use of nearly the amount of primitive neuroblasts—the embryonic and migrating nerve cells—possible in the lengthier, and at the same time paradoxically accelerated development of the human child. The clock in the body, in other words, has placed a limit upon the pace at which the ape brain grows—a limit which, as we have seen, the human ancestors in some manner escaped. This is a simplification of a complicated problem, but it hints at the answer to Wallace's question of long ago as to why man shows such a strange, rich mental life, many of whose artistic aspects can have had little direct value measured in the old utilitarian terms of the selection of all qualities in the struggle for existence.

When these released potentialities for brain growth began, they carried man into a new world where the old laws no longer totally held. With every advance in language, in symbolic thought, the brain paths multiplied. Significantly enough, those which are most heavily involved in the life processes, and are most ancient, mature first. The most recently acquired and less specialized regions of the brain, the "silent areas," mature last. Some neurologists, not without reason, suspect that here may lie other potentialities which only the future of the race may reveal.

Even now, however, the brain of man, with all its individual never-to-be-abandoned richness, is becoming merely a unit in the vast social brain which is potentially immortal, and whose memory is the heaped wisdom of the world's great thinkers. The scientist Haldane, brooding upon the future, has speculated that we will even further prolong our childhood and retard maturity if brain advance continues.

It is unlikely, however, in our present comfortable circumstances, that the pace of human change will ever again speed at the accelerated rate it knew when man strove against extinction. The story of Eden is a greater allegory than man has ever guessed. For it was truly man who, walking memoryless through bars of sunlight and shade in the morning of the world, sat down and passed a wondering hand across his heavy forehead. Time and darkness, knowledge of good and evil, have walked with him ever since. It is the destiny struck by the clock in the body in that brief space between the beginning of the first ice and that of the second. In just that interval a new world of terror and loneliness appears to have been created in the soul of man.

For the first time in four billion years a living creature had contemplated himself and heard with a sudden, unaccountable loneliness, the whisper of the wind in the night reeds. Perhaps he knew, there in the grass by the chill waters, that he had before him an immense journey. Perhaps that same foreboding still troubles the hearts of those who walk out of a crowded room and stare with relief into the abyss of space so long as there is a star to be seen twinkling across those miles of emptiness.

MAN OF THE FUTURE

There are days when I may find myself unduly pessimistic about the future of man. Indeed, I will confess that there have been occasions when I swore I would never again make the study of time a profession. My walls are lined with books expounding its mysteries, my hands have been split and raw with grubbing into the quicklime of its waste bins and hidden crevices. I have stared so much at death that I can recognize the lingering personalities in the faces of skulls and feel accompanying affinities and repulsions.

One such skull lies in the lockers of a great metropolitan museum. It is labeled simply: Strandlooper, South Africa. I have never looked longer into any human face than I have upon the features of that skull. I come there often, drawn in spite of myself. It is a face that would lend reality to the fantastic tales of our childhood. There is a hint of Wells's *Time Machine* folk in it—those pathetic, childlike people whom Wells pictures as haunting earth's autumnal cities in the far future of the dying planet.

Yet this skull has not been spirited back to us through future eras by a time machine. It is a thing, instead, of the millennial past. It is a caricature of modern man, not by reason of its primitiveness but, startlingly, because of a modernity outreaching his own. It constitutes, in fact, a mysterious prophecy and warning. For at the very moment in which students of humanity have been sketching their concept of the man of the future, that being has already come, and lived, and passed away.

We men of today are insatiably curious about ourselves and desperately in need of reassurance. Beneath our boisterous self-confidence is fear—a growing fear of the future we are in the process of creating. In such a mood we turn the pages of our favorite magazine and, like as not, come straight upon a description of the man of the future.

The descriptions are never pessimistic; they always, with sublime confidence, involve just one variety of mankind—our own—and they are always subtly flattering. In fact, a distinguished colleague of mine who was adept at this kind of prophecy once allowed a somewhat etherealized version of his own lofty brow to be used as an illustration of what the man of the future was to look like. Even the bald spot didn't matter—all the men of the future were to be bald, anyway.

Occasionally I show this picture to students. They find it highly comforting. Somebody with a lot of brains will save humanity at the proper moment. "It's all right," they say, looking at my friend's picture labeled "Man of the Future." "It's O.K. Somebody's keeping an eye on things. Our heads are getting bigger and our teeth are getting smaller. Look!"

Their voices ring with youthful confidence, the confidence engendered by my persuasive colleagues and myself. At times I glow a little with their reflected enthusiasm. I should like to regain that confidence, that warmth. I should like to but . . .

There's just one thing we haven't quite dared to mention. It's this, and you won't believe it. It's all happened already. Back there in the past, ten thousand years ago. The man of the future, with the big brain, the small teeth.

Where did it get him? Nowhere. *Maybe there isn't any future.* Or, if there is, maybe it's only what you can find in a little heap of bones on a certain South African beach.

Many of you who read this belong to the white race. We like to think about this man of the future as being white. It flatters our ego. But the man of the future in the past I'm talking about was not white. He lived in Africa. His brain was bigger than your brain. His face was straight and small, almost a child's face. He was the end evolutionary product in a direction quite similar to the one anthropologists tell us is the road down which we are traveling.

In the minds of many scholars, a process of "foetalization" is one of the chief mechanisms by which man of today has sloughed off his ferocious appearance of a million years ago, prolonged his childhood, and increased the size of his brain. "Foetalization" or "pedomorphism," as it is termed, means simply the retention, into adult life, of bodily characters which at some earlier stage of evolutionary history were actually only infantile. Such traits were rapidly lost as the animal attained maturity.

If we examine the life history of one of the existing great apes and compare its development with that of man, we observe that the infantile stages of both man and ape are far more similar than the two will be in maturity. At birth, as we have seen, the brain of the gorilla is close to the size of that of the human infant. Both newborn gorilla and human child are much more alike, facially, than they will ever be in adult life because the gorilla infant will, in the course of time, develop an enormously powerful and protrusive muzzle. The sutures of his skull will close early; his brain will grow very little more.

By contrast, human brain growth will first spurt and then grow steadily over an extended youth. Cranial sutures will remain open into adult life. Teeth will be later in their eruption. Furthermore, the great armored skull and the fighting characters of the anthropoid male will be held in abeyance.

Instead, the human child, through a more extended infancy, will approach a maturity marked by the retention of the smooth-browed skull of childhood. His jaws will be tucked inconspicuously under a forehead lacking the huge, muscle-bearing ridges of the ape. In some unknown manner, the ductless glands which stimulate or inhibit growth have, in the course of human evolution, stepped down the pace of development and increased the life span. Our helpless but well-cared-for childhood allows a longer time for brain growth and, as an indirect consequence, human development has slowly been steered away from the ape-like adulthood of our big-jawed forebears.

Modern man retains something of his youthful gaiety and nimble mental habits far into adult life. The great male anthropoids, by contrast, lose the playful friendliness of youth. In the end the massive skull houses a small, savage, and often morose brain. It is doubtful whether our thick-skulled forerunners viewed life very pleasantly in their advancing years.

We of today, then, are pedomorphs—the childlike, yet mature products of a simian line whose years have lengthened and whose adolescence has become long drawn out. We are, for our day and time, civilized. We eat soft food, and an Eskimo child can outbite us. We show signs, in our shortening jaws, of losing our wisdom teeth. Our brain has risen over our eyes and few, even of our professional fighters, show enough trace of a brow ridge to impress a half-grown gorilla. The signs point steadily onward toward a further lightening of the skull box and to additional compression of the jaws.

Imagine this trend continuing in modern man. Imagine our general average cranial capacity rising by two hundred cubic centimeters while the face continued to reduce proportionately. Obviously we would possess a much higher ratio of brain size to face size than now exists. We would, paradoxically, resemble somewhat our children of today. Children acquire facial prominence late in growth under the endocrine stimulus of maturity. Until that stimulus occurs, their faces bear a smaller ratio to the size of the brain case. It was so with these early South Africans.

But no, you may object, this whole process is in some way dependent upon civilization and grows out of it. Man's body and his culture mutually control each other. To that extent we are masters of our physical destiny. This mysterious change that is happening to our bodies is epitomized at just one point today, the point of the highest achieved civilization upon earth—our own.

I believed this statement once, believed it wholeheartedly. Sometimes it is so very logical I believe it still as my colleague's ascetic, earnest, and ennobled face gazes out at me from the screen. It carries the lineaments of my own kind, the race to which I belong. But it is not, I know now, the most foetalized race nor the largest brained. That game had already been played out before written history began—played out in an obscure backwater of the world where sails never came and where the human horde chipped flint as our ancestors had chipped it northward in Europe when the vast ice lay heavy on the land.

These people were not civilized; they were not white. But they meet in every major aspect the physical description of the man of tomorrow. They achieved that status on the raw and primitive diet of a savage. Their delicate and gracefully reduced teeth and fragile jaws are striking testimony to some strange inward hastening of change. Nothing about their environment in the least explains them. They were tomorrow's children surely, born by error into a lion country of spears and sand.

Africa is not a black man's continent in the way we are inclined to think. Like other great land areas it has its uneasy amalgams, its genetically strange variants, its racial deviants whose blood stream is no longer traceable. We know only that the first true men who disturbed the screaming sea birds over Table Bay were a folk that humanity has never looked upon again save as their type has wavered into brief emergence in an occasional mixed descendant. They are related in some dim manner to the modern Kalahari Bushman, but he is dwarfed in brain and body and hastening fast toward eventual extinction. The Bushman's forerunners, by contrast, might have stepped with Weena out of the future eras of the Time Machine.

Widespread along the South African coast, in the lowest strata of ancient cliff shelters, as well as inland in Ice Age gravel and other primeval deposits, lie the bones of these unique people. So remote are they from us in time that the first archaeologists who probed their caves and seashore middens had expected to reveal some distant and primitive human forerunner such as Neanderthal man. Instead their spades uncovered an unknown branch of humanity which, in the words of Sir Arthur Keith, the great English anatomist, "outrivals in brain volume any people of Europe, ancient or modern . . ."

But that is not all. Dr. Drennan of the University of Capetown comments upon one such specimen in anatomical wonder: "It appears ultramodern in many of its features, surpassing the European in almost every direction. That is to say, it is less simian than any modern skull." This ultramodernity Dr. Drennan attributes to the curious foetalization of which I have spoken.

More fascinating than big brain capacity in itself, however, is the relation of the cranium to the base of the skull and to the face. The skull base, that is, the part from the root of the nose to the spinal opening, is buckled and shortened in a way characteristic of the child's skull before the base expands to aid in the creation of the adult face. Thus, on this permanently shortened cranial base, the great brain expands, bulging the forehead heavily above the eyes and leaving the face neatly retracted beneath the brow. There is nothing in this face to suggest the protrusive facial angle of the true Negro. It is, as Dr. Drennan says, "ultramodern," even by Caucasian standards. The bottom of the skull grew, apparently, at a slow and childlike tempo while the pace-setting brain lengthened and broadened to a huge maturity.

When the skull is studied in projection and ratios computed, we find that these fossil South African folk, generally called "Boskop" or "Boskopoids" after the site of first discovery, have the amazing cranium-to-face ratio of almost five to one. In Europeans it is about three to one. This figure is a marked indication of the degree to which face size had been "modernized" and subordinated to brain growth. It is true that Dr. Ronald Singer has recently contended that the "Boskop" people cannot be successfully differentiated from the Bushman because Boskopoid features can be observed in this latter group, but even he would not deny the appearance of the peculiarly pedomorphic and ultrahuman features we have been discussing. At best, he would contend, in contrast to Keith and Drennan, that these characters have emerged in a sporadic fashion through-out the racial history of South Africa. By contrast, the facial structure of existing Caucasians, advanced though we imagine it, has only a mediocre rating.

The teeth vary a little from the usual idea about man of the future, yet they, too, are modern. Our prophecies generally include the speculation that we will, in time, lose our third molar teeth. This seems likely indeed, for the tooth often fails to erupt, crowds, and causes trouble. The Boskop folk had no such difficulty. Their teeth are small, neatly reduced in proportion to their delicate jaws, and free from any sign of the dental ills that trouble us. Here, in a hunter's world that would seem to have demanded at least the stout modern dentition of the Congo Negro, nature had decreed otherwise. These teeth could have nibbled sedately at the Waldorf, nor would the customers have been alarmed.

With the face, however, it would have been other-wise. In its anatomical structure we observe characters which relate these people both with the dwarf modern Bushman and to some ancient Negroid strain distinct from the West Coast blacks. We believe that they had the tightly-kinked "pepper-corn" hair of the Bushman as well as his yellow-brown skin. A branch of the Negro race has thus produced what is actually, so far as we can judge from the anatomical standpoint, one of the most ultrahuman types that ever lived! Had these characters appeared among whites, they would undoubtedly have been used in invidious comparisons with other "lesser" races.

We can, of course, repeat the final, unanswerable question: What did this tremendous brain mean to the Boskop people? We can marvel over their curious and exotic anatomy. We can wonder at the mysterious powers hidden in the human body, so potent that once unleashed they brought this more than modern being into existence on the very threshold of the Ice Age.

We can debate for days whether that magnificent cranial endowment actually represented a superior brain. We can smile pityingly at his miserable shell heaps, point to the mute stones that were his only tools. We can do this, but in doing it we are mocking our own rude forefathers of a similar day and time. We are forgetting the high artistic sensitivity which flowered in the closing Ice Age of Europe and which, oddly, blossomed here as well, lingering on even among the dwarfed Bushmen of the Kalahari. No, we cannot dismiss the Boskop people on such grounds, for even remarkable potential endowment cannot create high civilization overnight.

What we *can* say is that perhaps the unloosed mechanism ran too fast, that these people may have been ill-equipped physically to compete against the onrush of more ferocious and less foetalized folk. In a certain sense the biological clock had speeded them out of their time and place—a time which ten thousand years later has still not arrived. We may speculate that even mentally they may have lacked something of the elemental savagery of their competitors.

Their evolutionary gallop has led precisely nowhere save to a dwarfed and dying folk—if, with some authorities, we accept the later Bushmen as their descendants. This, then, was the logical end of complete foetalization: a desperate struggle to survive among a welter of more prolific and aggressive stocks. The answer to the one great question is still nowhere, still nothing. But there in the darkened laboratory, after the students have gone, I look once more at the exalted photograph of my friend upon the screen, noting character by character the foetalized refinement by which the artist has attempted to indicate the projected trend of future development—the expanded brain, the delicate face.

I look, and I know I have seen it all before, reading, as I have long grown used to doing, the bones through the living flesh. I have seen this face in another racial guise in another and forgotten day. And once again I grow aware of that eternal flickering of forms which we are now too worldly wise to label progress, and whose meaning forever escapes us.

The man of the future came, and looked out among us once with wistful, if unsophisticated eyes. He left his bones in the rubble of an alien land. If we read evolution aright, he may come again in another million years. Are the evolutionary forces searching for the right moment of his appearance? Or is his appearance itself destined always, even in the moment of emergence, to mark the end of the drama and foretell the extinction of a race?

Perhaps the strange interior clockwork that is here revealed as so indifferent to environmental surroundings has set, after all, a limit to the human time it keeps. That is the real question propounded by my friend's fine face. That is the question that I sometimes think the Boskop folk have answered. I wish I could be sure. I wish I knew.

Whatever else these skulls or those of occasional variant moderns may tell us, one thing they clearly reveal: Those who contend that because of present human cranial size, and the limitations of the human pelvis, man's brain is no longer capable of further expansion, are mistaken. Cranial capacities of almost a third more than the modern average have been occasionally attained among the Boskop people and even in rare individuals among other, less foetalized races. The secret does not lie in the size of the brain before birth; rather, as we have seen, it is contained in that strange spurt which in the first year of life carries man upward and outward into a social world from which his fellow beings are excluded. Whether that postnatal expansion is destined to be further enhanced in the long eras to come there is no telling, nor, perhaps, does it matter greatly. For in the creation of the social brain, nature, through man, has eluded the trap which has engulfed in one way or another every other form of life on the planet. Within the reasonable limits of the brain that now exists, she has placed the long continuity of civilized memory as it lies packed in the world's great libraries. The need is not really for more brains, the need is now for a gentler, a more tolerant people than those who won for us against the ice, the tiger, and the bear. The hand that hefted the ax, out of some old blind allegiance to the past fondles the machine gun as lovingly. It is a habit man will have to break to survive, but the roots go very deep.

I once sat, a prisoner, long ago, and watched a peasant soldier just recently equipped with a submachine gun swing the gun slowly into line with my body. It was a beautiful weapon and his finger toyed hesitantly with the trigger. Suddenly to possess all that power and then to be forbidden to use it must have been almost too much for the man to contain. I remember, also, a protesting female voice nearby—the eternal civilizing voice of women who know that men are fools and children, and irresponsible. Sheepishly the peon slowly dropped the gun muzzle away from my chest. The black eyes over the barrel looked out at me a little wicked, a little desirous of better understanding.

"Thompson, Tome'-son'," he repeated proudly,

slapping the barrel. "Tome'-son'." I nodded a little weakly, relaxing with a sigh. After all, we were men together and understood this great subject of destruction. And was I not a citizen of the country that had produced this wonderful mechanism? So I nodded again and said carefully after him. "Thompson, Tome'-son'. *Bueno, si, muy bueno.*" We looked at each other then, smiling a male smile that ran all the way back to the Ice Age. In academic halls since, considering the future of humanity, I have never been quite free of the memory of that soldier's smile. I weigh it mentally against the future whenever one of those delicate forgotten skulls is placed upon my desk.

LITTLE MEN
AND FLYING SAUCERS

✤

Today, as never before, the sky is menacing. Things
seen indifferently last century by the wandering lamp-
lighter now trouble a generation that has grown up to
the wail of air-raid sirens and the ominous expectation
that the roof may fall at any moment. Even in daytime,
reflected light on a floating dandelion seed, or a spider
riding a wisp of gossamer in the sun's eye, can bring
excited questions from the novice unused to estimating
the distance or nature of aerial objects.

Since we now talk, write, and dream endlessly of
space rockets, it is no surprise that this thinking yields

the obverse of the coin: that the rocket or its equivalent may have come first to us from somewhere "outside." As a youth, I may as well confess, I waited expectantly for it to happen. So deep is the conviction that there must be life out there beyond the dark, one thinks that if they are more advanced than ourselves they may come across space at any moment, perhaps in our generation. Later, contemplating the infinity of time, one wonders if perchance their messages came long ago, hurtling into the swamp muck of the steaming coal forests, the bright projectile clambered over by hissing reptiles, and the delicate instruments running mindlessly down with no report.

Sometimes when young, and fossil hunting in the western Badlands, I had thought it might yet be found, corroding and long dead, in the Tertiary sod that was once green under the rumbling feet of titanotheres. Surely, in the infinite wastes of time, in the lapse of suns and wane of systems, the passage, if it were possible, would have been achieved. But the bright projectile has not been found and now, in sobering middle age, I have long since ceased to look. Moreover, the present theory of the expanding universe has made time, as we know it, no longer infinite. If the entire universe was created in a single explosive instant a few billion years ago, there has not been a sufficient period for all things to occur even behind the star shoals of the outer galaxies. In the light of this fact it is now just conceivable that there may be nowhere in space a mind superior to our own.

If such a mind should exist, there are many reasons why it could not reside in the person of a little man. There is, however, a terrible human fascination about the miniature, and one little man in the hands of the spinner of folk tales can multiply with incredible rapidity. Our unexplainable passion for the small is not quenched at the borders of space, nor, as we shall see, in the spinning rings of the atom. The flying saucer and the much publicized little men from space equate neatly with our own projected dreams.

When I first heard of the little man there was no talk of flying saucers, nor did his owner ascribe to him anything more than an earthly origin. It has been almost a quarter of a century since I encountered him in a bone hunter's camp in the West. A rancher had brought him to us in a box. "I figured you'd maybe know about him," he said. "He'll cost you money, though. There's money in that little man."

"Man?" we said.

"Man," he countered. "What you'd call a pygmy or a dwarf, but smaller than any show dwarf I ever did see. A mummy, too, a little dead mummy. I figure it was some kind of bein' like us, but little. They put him in the place I found him; maybe it was a thousand years ago. You'll likely know."

Our heads met over the box. The last paper was withdrawn. The creature emerged on the man's palm. I've seen a lot of odd things in the years since, and fakes by the score, but that little fellow gave me the creeps. He might have been two feet high in a standing posture —not more. He was mummified in a crouching position, arms folded. The face with closed eyes seemed vaguely evil. I could have sworn I was dreaming.

I touched it. There was a peculiar, fleshy consistency about it, still. It was not a dry mummy. It was more like what you would expect a natural cave mummy to be like. It had no tail. I know because I looked. And to this day the little man sits on there, in my brain, and as plain as yesterday I can see the faint half-smirk of his mouth and the tiny black hands at his knees.

"You can have it for two hundred bucks," said the man. We glanced at each other, sighed, and shook our heads. "We aren't in the market," we said. "We're collecting, not buying, and we're staying with our bones."

"Okay," said the man, and gave us a straight look, closing his box. "I'm going to the carnival down below tonight. There's money in him. There's money in that little man."

I think it may have been just as well for us that we made no purchase. I have never liked the little man, nor the description of the carnival to which he and his owner were going. It may be, I used to think, that I will yet encounter him before I die, in some little colored tent on a country midway. Once, in the years since, I have heard a description that sounded like him in another guise. It involved a fantastic tale of some Paleozoic beings who hunted among the tree ferns when the world was ruled by croaking amphibians. The story did not impress me; I knew him by then for what he was: an anomalous mummified stillbirth with an undeveloped brain.

I never expected to see him emerge again in books on flying saucers, or to see the "little men" multiply and become so common that columnists would take note of them. Nor, though I should have known better, did I expect to live to hear my little man ascribed an extraplanetary origin. There is a story back of him, it is true, but it is a history of this earth, and, of all unlikely things, it involves that great man of science, Charles Darwin, though by a curious, lengthy, and involved route.

Men have been men for so long that they tend not to question the fact. All their experience tells them that their children will precisely resemble themselves; that kittens will become cats and cats will have kittens, and that even caterpillars, though the pattern seems a little odd, will become butterflies, and butterflies will produce caterpillars. It is so habitual an event that we do not stop to ask why this happens, or to consider that this amazing precision in results implies a strange order ing of life in a world we often think is chanceful and meaningless.

A few wise men since the time of the Greeks have found it a source of wonder, but they have been a minority. Most people have shrugged and spoken indifferently of the gods, or contented themselves, as the Christian world did for so long, with the idea of special creation of each species. Nevertheless, the wise ones kept on wondering.

They found, as they began their first groping attempts to classify and arrange the living world, that in spite of the assumed individual creation of every living species by the supernatural intervention of divine power, a basic similarity of structure existed among many forms of life. This was a remarkable thing to find among supposedly individual creations. Offhand one would say that a much greater degree of spontaneous novelty would have been possible. In fact, man once innocently believed himself part of such a creation. The fabulous animals of the ancient bestiaries, the mermaids, griffins, and centaurs, not to mention the men whose ears were so large that their owners slept in them, would have been the natural, spontaneous products of such uncontrolled, creative whimsy.

But there was the pattern: the ape and the man with their bone-by-bone correspondence. The very fact that one can add a plural to the word *reptile* and so suggest anything from a brontosaurus to a garter snake shows that a pattern exists. Birds all have feathers, wings, and claws; they are a common class in spite of their diversities. They have been pulled into many shapes, but there is still an eternal "birdliness" about them. They are built on a common plan, just as I share mammalian characters with a small mouse who inhabits my desk drawer. This is hard to account for in a disordered world, so that recently, when I came upon this mouse, trapped and terrified in the wastebasket, his similarity to myself rendered me helpless, and out of sheer embarrassment I connived in his escape.

Now so long as these remarkable patterns could be observed only in the living world around us, they occasioned no great alarm. Even after Cuvier, in 1812, made a magnificent attempt to reduce the forms of animal life to four basic blueprints or "archetypes" of divergent character, no one was particularly disturbed—least of all from the religious point of view. In the words of one great naturalist, Louis Agassiz, "This plan of creation . . . has not grown out of the necessary action of physical laws, but was the free conception of the Almighty Intellect, matured in his thought before it was manifested in tangible external forms."

It was not long, however, before *pattern*, the divine blueprint, first recognized in the existing world, was extended by the geologist across the deeps of time. The animal world of the past was in the process of discovery. It proved to be a world without man. Curiously enough, it was soon learned that extinct animals could be fitted into the broad classifications of the existing world. They were mammals or amphibia or reptiles, as the case might be. Though no living eye had beheld them, they seemed to mark the continuation of the divine abstraction, the eternal patterns, across the enormous time gulfs of the past.

The second fact, that man had not been discovered, was a cause for dismay. In the man-centered universe of the time, one can appreciate the anguish of the Reverend Mr. Kirby discovering the Age of Reptiles: "Who can think that a being of unbounded power, wisdom, and goodness, should create a world merely for the habitation of a race of monsters, without a single, rational being in it to serve and glorify him?" This is the wounded outcry of the human ego as it fails to discover its dominance among the beasts of the past. Even more tragically, it learns that the world supposedly made for its enjoyment has existed for untold eons entirely indifferent to its coming. The chill vapors of time and space are beginning to filter under the closed door of the human intellect.

It was in these difficult straits, in the black night of his direst foreboding, that the doctrine of geologic prophecy was evolved by man. For fifty years it would hold time at bay, and in one last great effort its proponents, by clever analogies, would attempt to extend the human drama across the infinite worlds of space; it echoes among us still in the shape of the little men of the flying saucers. No braver *mythos* was ever devised under the cold eye of science.

In an old book from my shelves, Hugh Miller's *The Testimony of the Rocks*, I find this passage: "*Higher still in one of the deposits of the Trias we are startled by what seems to be the impression of a human hand of an uncouth massive shape, but with the thumb apparently set in opposition, as in man, to the other fingers.*"

There is only one way to understand this literature. The biologists of the first half of the nineteenth century had recognized that the unity of animal organization descends into past ages and is observable in forms no living eye has beheld. It was, they believed, an immaterial, a supernatural line of connection. They refused to see in this unity of plan an actual physical relationship. Instead they read the past as a successive series of creations and extinctions upon a divinely modifiable but consistent plan. "Geology," said one writer, "unrolls a prophetic scroll, in which the earlier animated creation points on to the later."

In 1726, before the rise of geological theology, Professor Scheuchzer of Zurich had discovered and described the skeleton of a long extinct amphibian as that of *Homo Diluvii testis*, "Man, witness of the flood." The remains, after being piously termed "a rare relic of the accursed race of the primitive world," were found to be those of an animal, and interest in the fossil ceased. With the development of geological prophecy, however, we find this giant salamander reappearing in the writings of that eminent Scotch philosopher, James McCosh. Admitting the true nature of the relic, McCosh, undaunted, contended in 1857: "Long ages had yet to roll on before the consummation of the vertebrate type; the preparations for man's appearance were not yet completed. *Nevertheless, in this fossil of Scheuchzer's there was a prefiguration of the more perfect type which man's bony framework presents.*" Thus the swinging pick of the geologist at work in the world's bone yards did not, at first, disturb the abstract beauty of the Platonic forms. Instead, the recognition of the past enveloped life with a strange premonitory quality, a sense of prophecy and doom as carefully ordered as the movement on some great stage.

It is in the light of this philosophy that the hand, "massive" and of "uncouth shape," must be interpreted. It foreshadows, out of that slimy concourse of sprawling amphibians and gaping lizards, the eventual emergence of man. Splayed, monstrous, and mud-smeared, it haunts the future. That it is the footprint of some wandering reptilian beast of the coal swamps may be granted, but it is also a vertebrate. Its very body forecasts the times to come.

It would be erroneous, however, to conceive of reptiles as being the major preoccupation of our geological prophets. They scanned the anatomy of fishes, birds, and salamanders, seeking in their skeletons anticipations of the more perfect structure of man. If they found footprints of fossil bipeds it was a "sign" foretelling man. All things led in his direction. Prior to his entrance the stage was merely under preparation. In this way the blow to the human ego had been softened. The past was only the prologue to the Great Play. Man was at the heart of things after all.

It was a strange half century, as one looks back upon it—that fifty years before the publication of Darwin's *Origin of Species*. It was dominated by a generation that saw the world as a complex symbolic system pointing in the direction of man, who was foreknown and prefigured from the beginning. Man, who comes last, is the end of this strange cycle. With him, in the eyes of many of these thinkers, the process ceases and no further changes in the world of life are to be expected. Since the transcendental "evolutionists" were man-centered, questions involving divergent evolution and adaptation did not come easily to their minds. Working with an immaterial and abstract Platonic concept, it was inevitable that they should seek to extend their doctrine across the deeps of space. Because the pattern was capable of modification, the possibility of the existence of small men, large men, or men of different colors upon other planets did not trouble them, but men they ought to be. There was little comprehension of the fact that man had acquired his particular bodily structure and upright posture through a peculiar set of evolutionary circumstances, not easily to be duplicated.

The theory of the plurality of worlds is a very ancient one; that is, the notion that the lights seen elsewhere in space may be bodies like that which we inhabit. After the rise of the Copernican astronomy and the growing realization that our earth is part of a planetary system revolving around a central sun, it was often contended by philosophers that the other stars seen in space must be similar suns with similar planetary satellites.

Quarrels arose between those who believed God's power infinitely and creatively extended among the stars, and those who regarded it as heresy and dangerous to Christian belief to imply that the Infinite Mind might be concerned with more than the beings of this planet. It was a struggle heightened by an enormous extension of man's vision into the worlds of the infinitely far and the infinitely small, the telescope and the microscope having momentarily stunned the human imagination. Some clung frantically to the little tight-fenced world of the Middle Ages, refusing to acknowledge what these instruments revealed. Others, with greater willingness to accept the new, tried, nevertheless, to equate what they saw with old beliefs and to elaborate an "astrotheology."

In the fifties of the last century there was a great outburst of interest in the possibility of life on other worlds. The recently discovered life history of our own planet and improvements in astronomical apparatus had all excited great interest on the part of a public wavering in its loyalty between old religious dogmas and the new revelations of science. Speculation, in many instances, was roaming far in advance of actual observation.

"The inhabitants of Jupiter," wrote William Whewell in 1854, "must . . . it would seem, be cartilaginous and glutinous masses. If life be there it does not seem in any way likely, that the living things can be anything higher in the scale of being, than such boneless, watery, pulpy creatures . . ."

This remark is not intended as merely innocent theorizing. In his work, *Plurality of Worlds*, Whewell indicates his definite opposition to the idea that the other planets, or the more remote worlds in other galaxies, are inhabited. At best he is willing to grant the existence of a few gelatinous creatures such as he mentions in the above passage, but that man is to be found elsewhere, he denies. He argues that there are superior and inferior regions of space. Man, preceded by endless eons of lower creatures in time, is yet a superior being. He calls attention to the fact that "the intelligent part of creation is thrust into the compass of a few years, in the course of myriads of ages; why not then into the compass of a few miles, in the expanse of systems?" **On** this earth a "supernatural interposition" has introduced man; the planet is unique.

Whewell's essay generated a storm of discussion. His was not the popular side of the controversy. Sir David Brewster countered with a volume significantly titled *More Worlds Than One*, in which he bluntly asserts: "The function of one satellite must be the function of all the rest. The function of our Moon, to give light to the Earth, must be the function of the other twenty-two moons of the system; and the function of the Earth, *to support inhabitants*, must be the function of all other planets." He dwells on the "grand combination" of "*infinity of life, with infinity of matter.*"

Brewster, moreover, calls attention to the invisible domain revealed by the microscope and argues from this that God has all along been attentive to forms of life of which we had no knowledge. So intriguing became the relativity of size that one author even produced a work whose subtitle bore the query *Are Ultimate Atoms Inhabited Worlds?* Stories like Fitz-James O'Brien's "The Diamond Lens," or Ray Cummings' "The Girl in the Golden Atom," stem from such thought.

Another writer, William Williams, in *The Universe No Desert, the Earth No Monopoly*, strikes more directly at the heart of the argument. He invokes geological prophecy and extends it directly across space:

"The archetypal idea of man, revealed in the lower vertebrated animals, proves God's foreknowledge of man's existence; and it equally applies to vertebrates on Jupiter or Neptune as to those on the Earth; and still farther, to the Universe, as these animals were within its precincts."

Williams was not the first nor the last man to utter these sentiments, but he did so with a fierce singleness of purpose. The life plans were immanent, prophetic, and immaterial. They could thus be projected across space. Why, he argues with the same horror that the Reverend Mr. Kirby had exhibited toward the Age of Reptiles, should God "banish his own image to one diminutive enclosure and surround . . . the residue of His immense Person with unintelligent, half-formed, crude monsters?" If man is regarded as a good production here, he must be found in endless duplication throughout the worlds. The pattern in the rocks of this earth is the pattern of the whole.

The shattering of this scheme of geological prophecy was the work of many men, but it was Charles Darwin who brought the event to pass, and who engineered what was to be one of the most dreadful blows that the human ego has ever sustained: the demonstration of man's physical relationship to the world of the lower animals. It is quite apparent, however, that there is an aspect of Darwin's discoveries which has never penetrated to the mind of the general public. It is the fact that once undirected variation and natural selection are introduced as the mechanism controlling the development of plants and animals, the evolution of every world in space becomes a series of unique historical events. The precise accidental duplication of a complex form of life is extremely unlikely to occur in even the same environment, let alone in the different background and atmosphere of a far-off world.

In the modern literature on space travel I have read about cabbage men and bird men; I have investigated the loves of the lizard men and the tree men, but in each case I have labored under no illusion. I have been reading about a man, *Homo sapiens*, that common earthling, clapped into an ill-fitting coat of feathers and retaining all his basic human attributes including an eye for the pretty girl who has just emerged from the space ship. His lechery and miscegenating proclivities have an oddly human ring, and if this is all we are going to find on other planets, I, for one, am going to be content to stay at home. There is quite enough of that sort of thing down here, without encouraging it throughout the starry systems.

The truth is that man is a solitary and peculiar development. I do not mean this in any irreverent or contemptuous sense. I want merely to point out that when Charles Darwin and his colleagues established the community of descent of the living world, and observed the fact of divergent evolutionary adaptation, they destroyed forever the concept of geological prophecy. They did not eliminate the possibility of life on other worlds, but the biological principles which they established have totally removed the likelihood that our descendants, in the next few decades, will be entertaining little men from Mars. I would be much more willing to consider the possibility of sitting down to lunch with a purple polyp, but even this has anatomical comparisons with the life of this planet.

Geologic prophecy was based on two things: first, a belief, as we have seen, in the man-centered nature of the universe, and second, the assumption that since the animals of the past had no physical connection with those of the present, some kind of abstract, immaterial plan in the mind of the Creator linked the forms of the past with those of the present day. The early-nineteenth-century thinkers perceived a genuine relationship, but their attachment to the idea of special creation prevented them from recognizing that the relationship arose out of simple biological "descent with modification."

Man could not be proved preordained or predestined from the beginning simply because he showed certain affinities to Paleozoic vertebrates. Instead, he was merely one of many descendants of the early vertebrate line. A moose or a mongoose would have had equally good reason to contend that as a modern vertebrate he had been "prefigured from the beginning," and that the universe had been organized with him in mind.

The situation is something like that of walking through a hall of trick mirrors and being pulled out of shape. The mirror of time does that to all things living, and the distortions stay. Nevertheless, there is a pattern of sorts, so that if you have come by the mirror that makes men, and somewhere behind you there is a mirror that makes black cats, you can still see the pattern. You and the cat are related; the shreds of the original shape are in your bones and the shreds of primordial thought patterns move in the eyes of both of you and are understood by both. But somewhere there must be an original pattern; somewhere cat and man and weasel must leap into a single shape. That shape lies inconceivably remote from us now, far back along the time stream. It is historical. In that sense, and in that sense only, the archetype did indeed exist.

Darwin saw clearly that the succession of life on this planet was not a formal pattern imposed from without, or moving exclusively in one direction. Whatever else life might be, it was adjustable and not fixed. It worked its way through difficult environments. It modified and then, if necessary, it modified again, along roads which would never be retraced. Every creature alive is the product of a unique history. The statistical probability of its precise reduplication on another planet is so small as to be meaningless. Life, even cellular life, may exist out yonder in the dark. But high or low in nature, it will not wear the shape of man. That shape is the evolutionary product of a strange, long wandering through the attics of the forest roof, and so great are the chances of failure, that nothing precisely and identically human is likely ever to come that way again.

The picture of the little man of long ago rises before me as I write. As I have said, he was simply a foetal monster, long since scientifically diagnosed and dismissed. The small skull that lent the illusion of maturity to the mummified infant contained a brain which had failed to develop. The describers of two-foot men forget that a normal human brain cannot function with a capacity, at the very minimum, of less than about nine hundred cubic centimeters. A man with a hundred-cubic-centimeter brain will not be a builder of flying saucers; he will be less intelligent than an ape. In any case, he does not exist.

In a universe whose size is beyond human imagining, where our world floats like a dust mote in the void of night, men have grown inconceivably lonely. We scan the time scale and the mechanisms of life itself for portents and signs of the invisible. As the only thinking mammals on the planet—perhaps the only thinking animals in the entire sidereal universe—the burden of consciousness has grown heavy upon us. We watch the stars, but the signs are uncertain. We uncover the bones of the past and seek for our origins. There is a path there, but it appears to wander. The vagaries of the road may have a meaning, however; it is thus we torture ourselves.

Lights come and go in the night sky. Men, troubled at last by the things they build, may toss in their sleep and dream bad dreams, or lie awake while the meteors whisper greenly overhead. But nowhere in all space or on a thousand worlds will there be men to share our loneliness. There may be wisdom; there may be power; somewhere across space great instruments, handled by strange, manipulative organs, may stare vainly at our floating cloud wrack, their owners yearning as we yearn. Nevertheless, in the nature of life and in the principles of evolution we have had our answer. Of men elsewhere, and beyond, there will be none forever.

THE JUDGMENT
OF THE BIRDS

It is a commonplace of all religious thought, even the most primitive, that the man seeking visions and insight must go apart from his fellows and live for a time in the wilderness. If he is of the proper sort, he will return with a message. It may not be a message from the god he set out to seek, but even if he has failed in that particular, he will have had a vision or seen a marvel, and these are always worth listening to and thinking about.

The world, I have come to believe, is a very queer place, but we have been part of this queerness for so long that we tend to take it for granted. We rush to and fro like Mad Hatters upon our peculiar errands, all the time imagining our surroundings to be dull and ourselves quite ordinary creatures. Actually, there is nothing in the world to encourage this idea, but such is the mind of man, and this is why he finds it necessary from time to time to send emissaries into the wilderness in the hope of learning of great events, or plans in store for him, that will resuscitate his waning taste for life. His great news services, his world-wide radio network, he knows with a last remnant of healthy distrust will be of no use to him in this matter. No miracle can withstand a radio broadcast, and it is certain that it would be no miracle if it could. One must seek, then, what only the solitary approach can give—a natural revelation.

Let it be understood that I am not the sort of man to whom is entrusted direct knowledge of great events or prophecies. A naturalist, however, spends much of his life alone, and my life is no exception. Even in New York City there are patches of wilderness, and a man by himself is bound to undergo certain experiences falling into the class of which I speak. I set mine down, therefore: a matter of pigeons, a flight of chemicals, and a judgment of birds, in the hope that they will come to the eye of those who have retained a true taste for the marvelous, and who are capable of discerning in the flow of ordinary events the point at which the mundane world gives way to quite another dimension.

New York is not, on the whole, the best place to enjoy the downright miraculous nature of the planet. There are, I do not doubt, many remarkable stories to be heard there and many strange sights to be seen, but to grasp a marvel fully it must be savored from all aspects. This cannot be done while one is being jostled and hustled along a crowded street. Nevertheless, in any city there are true wildernesses where a man can be alone. It can happen in a hotel room, or on the high roofs at dawn.

One night on the twentieth floor of a midtown hotel I awoke in the dark and grew restless. On an impulse I climbed upon the broad old-fashioned window sill, opened the curtains and peered out. It was the hour just before dawn, the hour when men sigh in their sleep, or, if awake, strive to focus their wavering eyesight upon a world emerging from the shadows. I leaned out sleepily through the open window. I had expected depths, but not the sight I saw.

I found I was looking down from that great height into a series of curious cupolas or lofts that I could just barely make out in the darkness. As I looked, the outlines of these lofts became more distinct because the light was being reflected from the wings of pigeons who, in utter silence, were beginning to float outward upon the city. In and out through the open slits in the cupolas passed the white-winged birds on their mysterious errands. At this hour the city was theirs, and quietly, without the brush of a single wing tip against stone in that high, eerie place, they were taking over the spires of Manhattan. They were pouring upward in a light that was not yet perceptible to human eyes, while far down in the black darkness of the alleys it was still midnight.

As I crouched half asleep across the sill, I had a moment's illusion that the world had changed in the night, as in some immense snowfall, and that if I were to leave, it would have to be as these other inhabitants were doing, by the window. I should have to launch out into that great bottomless void with the simple confidence of young birds reared high up there among the familiar chimney pots and interposed horrors of the abyss.

I leaned farther out. To and fro went the white wings, to and fro. There were no sounds from any of them. They knew man was asleep and this light for a little while was theirs. Or perhaps I had only dreamed about man in this city of wings—which he could surely never have built. Perhaps I, myself, was one of these birds dreaming unpleasantly a moment of old dangers far below as I teetered on a window ledge.

Around and around went the wings. It needed only a little courage, only a little shove from the window ledge to enter that city of light. The muscles of my hands were already making little premonitory lunges. I wanted to enter that city and go away over the roofs in the first dawn. I wanted to enter it so badly that

I drew back carefully into the room and opened the hall door. I found my coat on the chair, and it slowly became clear to me that there was a way down through the floors, that I was, after all, only a man.

I dressed then and went back to my own kind, and I have been rather more than usually careful ever since not to look into the city of light. I had seen, just once, man's greatest creation from a strange inverted angle, and it was not really his at all. I will never forget how those wings went round and round, and how, by the merest pressure of the fingers and a feeling for air, one might go away over the roofs. It is a knowledge, however, that is better kept to oneself. I think of it sometimes in such a way that the wings, beginning far down in the black depths of the mind, begin to rise and whirl till all the mind is lit by their spinning, and there is a sense of things passing away, but lightly, as a wing might veer over an obstacle.

To see from an inverted angle, however, is not a gift allotted merely to the human imagination. I have come to suspect that within their degree it is sensed by animals, though perhaps as rarely as among men. The time has to be right; one has to be, by chance or intention, upon the border of two worlds. And sometimes these two borders may shift or interpenetrate and one sees the miraculous.

I once saw this happen to a crow.

This crow lives near my house, and though I have

never injured him, he takes good care to stay up in the very highest trees and, in general, to avoid humanity. His world begins at about the limit of my eyesight.

On the particular morning when this episode occurred, the whole countryside was buried in one of the thickest fogs in years. The ceiling was absolutely zero. All planes were grounded, and even a pedestrian could hardly see his outstretched hand before him.

I was groping across a field in the general direction of the railroad station, following a dimly outlined path. Suddenly out of the fog, at about the level of my eyes, and so closely that I flinched, there flashed a pair of immense black wings and a huge beak. The whole bird rushed over my head with a frantic cawing outcry of such hideous terror as I have never heard in a crow's voice before, and never expect to hear again.

He was lost and startled, I thought, as I recovered my poise. He ought not to have flown out in this fog. He'd knock his silly brains out.

All afternoon that great awkward cry rang in my head. Merely being lost in a fog seemed scarcely to account for it—especially in a tough, intelligent old bandit such as I knew that particular crow to be. I even looked once in the mirror to see what it might be about me that had so revolted him that he had cried out in protest to the very stones.

Finally, as I worked my way homeward along the path, the solution came to me. It should have been clear before. The borders of our worlds had shifted. It was the fog that had done it. That crow, and I knew him well, never under normal circumstances flew low near men. He had been lost all right, but it was more than that. He had thought he was high up, and when he encountered me looming gigantically through the fog, he had perceived a ghastly and, to the crow mind, unnatural sight. He had seen a man walking on air, desecrating the very heart of the crow kingdom, a harbinger of the most profound evil a crow mind could conceive of—air-walking men. The encounter, he must have thought, had taken place a hundred feet over the roofs.

He caws now when he sees me leaving for the station in the morning, and I fancy that in that note I catch the uncertainty of a mind that has come to know things are not always what they seem. He has seen a marvel in his heights of air and is no longer as other crows. He has experienced the human world from an unlikely perspective. He and I share a viewpoint in common: our worlds have interpenetrated, and we both have faith in the miraculous.

It is a faith that in my own case has been augmented by two remarkable sights. As I have hinted previously, I once saw some very odd chemicals fly across a waste so dead it might have been upon the moon, and once, by an even more fantastic piece of luck, I was present when a group of birds passed a judgment upon life.

On the maps of the old voyageurs it is called *Mauvaises Terres*, the evil lands, and, slurred a little with the passage through many minds, it has come down to us anglicized as the Badlands. The soft shuffle of moccasins has passed through its canyons on the grim business of war and flight, but the last of those slight disturbances of immemorial silences died out almost a century ago. The land, if one can call it a land, is a waste as lifeless as that valley in which lie the kings of Egypt. Like the Valley of the Kings, it is a mausoleum, a place of dry bones in what once was a place of life. Now it has silences as deep as those in the moon's airless chasms.

Nothing grows among its pinnacles; there is no shade except under great toadstools of sandstone whose bases have been eaten to the shape of wine glasses by the wind. Everything is flaking, cracking, disintegrating, wearing away in the long, imperceptible weather of time. The ash of ancient volcanic outbursts still sterilizes its soil, and its colors in that waste are the colors that flame in the lonely sunsets on dead planets. Men come there but rarely, and for one purpose only, the collection of bones.

It was a late hour on a cold, wind-bitten autumn day when I climbed a great hill spined like a dinosaur's back and tried to take my bearings. The tumbled waste fell away in waves in all directions. Blue air was darkening into purple along the bases of the hills. I shifted my knapsack, heavy with the petrified bones of long-vanished creatures, and studied my compass. I wanted to be out of there by nightfall, and already the sun was going sullenly down in the west.

It was then that I saw the flight coming on. It was moving like a little close-knit body of black specks that danced and darted and closed again. It was pouring from the north and heading toward me with the undeviating relentlessness of a compass needle. It streamed through the shadows rising out of monstrous gorges. It rushed over towering pinnacles in the red light of the sun, or momentarily sank from sight within their shade. Across that desert of eroding clay and wind-worn stone they came with a faint wild twittering that filled all the air about me as those tiny living bullets hurtled past into the night.

It may not strike you as a marvel. It would not, perhaps, unless you stood in the middle of a dead world at sunset, but that was where I stood. Fifty million years lay under my feet, fifty million years of bellowing monsters moving in a green world now gone so utterly that its very light was travelling on the farther edge of space. The chemicals of all that vanished age lay about me in the ground. Around me still lay the shearing molars of dead titanotheres, the delicate sabers of soft-stepping cats, the hollow sockets that had held the eyes of many a strange, outmoded beast. Those eyes had looked out upon a world as real as ours; dark, savage brains had roamed and roared their challenges into the steaming night.

Now they were still here, or, put it as you will, the chemicals that made them were here about me in the ground. The carbon that had driven them ran blackly in the eroding stone. The stain of iron was in the clays. The iron did not remember the blood it had once moved within, the phosphorus had forgot the savage brain. The little individual moment had ebbed from all those strange combinations of chemicals as it would ebb from our living bodies into the sinks and runnels of oncoming time.

I had lifted up a fistful of that ground. I held it while that wild flight of south-bound warblers hurtled over me into the oncoming dark. There went phosphorus, there went iron, there went carbon, there beat the calcium in those hurrying wings. Alone on a dead planet I watched that incredible miracle speeding past. It ran by some true compass over field and waste land. It cried its individual ecstasies into the air until the gullies rang. It swerved like a single body, it knew itself and, lonely, it bunched close in the racing darkness, its individual entities feeling about them the rising night. And so, crying to each other their identity, they passed away out of my view.

I dropped my fistful of earth. I heard it roll inanimate

back into the gully at the base of the hill: iron, carbon, the chemicals of life. Like men from those wild tribes who had haunted these hills before me seeking visions, I made my sign to the great darkness. It was not a mocking sign, and I was not mocked. As I walked into my camp late that night, one man, rousing from his blankets beside the fire, asked sleepily, "What did you see?"

"I think, a miracle," I said softly, but I said it to myself. Behind me that vast waste began to glow under the rising moon.

I have said that I saw a judgment upon life, and that it was not passed by men. Those who stare at birds in cages or who test minds by their closeness to our own may not care for it. It comes from far away out of my past, in a place of pouring waters and green leaves. I shall never see an episode like it again if I live to be a hundred, nor do I think that one man in a million has ever seen it, because man is an intruder into such silences. The light must be right, and the observer must remain unseen. No man sets up such an experiment. What he sees, he sees by chance.

You may put it that I had come over a mountain, that I had slogged through fern and pine needles for half a long day, and that on the edge of a little glade with one long, crooked branch extending across it, I had sat down to rest with my back against a stump. Through accident I was concealed from the glade, although I could see into it perfectly.

The sun was warm there, and the murmurs of forest life blurred softly away into my sleep. When I awoke, dimly aware of some commotion and outcry in the clearing, the light was slanting down through the pines in such a way that the glade was lit like some vast cathedral. I could see the dust motes of wood pollen in the long shaft of light, and there on the extended branch sat an enormous raven with a red and squirming nestling in his beak.

The sound that awoke me was the outraged cries of the nestling's parents, who flew helplessly in circles about the clearing. The sleek black monster was indifferent to them. He gulped, whetted his beak on the dead branch a moment and sat still. Up to that point the little tragedy had followed the usual pattern. But suddenly, out of all that area of woodland, a soft sound of complaint began to rise. Into the glade fluttered small birds of half a dozen varieties drawn by the anguished outcries of the tiny parents.

No one dared to attack the raven. But they cried there in some instinctive common misery, the bereaved and the unbereaved. The glade filled with their soft rustling and their cries. They fluttered as though to point their wings at the murderer. There was a dim intangible ethic he had violated, that they knew. He was a bird of death.

And he, the murderer, the black bird at the heart of life, sat on there, glistening in the common light, formidable, unmoving, unperturbed, untouchable.

The sighing died. It was then I saw the judgment. It was the judgment of life against death. I will never see it again so forcefully presented. I will never hear it again in notes so tragically prolonged. For in the midst of protest, they forgot the violence. There, in that clearing, the crystal note of a song sparrow lifted hesitantly in the hush. And finally, after painful fluttering, another took the song, and then another, the song passing from one bird to another, doubtfully at first, as though some evil thing were being slowly forgotten. Till suddenly they took heart and sang from many throats joyously together as birds are known to sing. They sang because life is sweet and sunlight beautiful. They sang under the brooding shadow of the raven. In simple truth they had forgotten the raven, for they were the singers of life, and not of death.

I was not of that airy company. My limbs were the heavy limbs of an earthbound creature who could climb mountains, even the mountains of the mind, only by a great effort of will. I knew I had seen a marvel and observed a judgment, but the mind which was my human endowment was sure to question it and to be at me day by day with its heresies until I grew to doubt the meaning of what I had seen. Eventually darkness and subtleties would ring me round once more.

And so it proved until, on the top of a stepladder, I made one more observation upon life. It was cold that autumn evening, and, standing under a suburban street light in a spate of leaves and beginning snow, I was suddenly conscious of some huge and hairy shadows dancing over the pavement. They seemed attached to an odd, globular shape that was magnified above me. There was no mistaking it. I was standing under the shadow of an orb-weaving spider. Gigantically projected against the street, she was about her spinning when everything was going underground. Even her cables were magnified upon the sidewalk and already I was half-entangled in their shadows.

"Good Lord," I thought, "she has found herself a kind of minor sun and is going to upset the course of nature."

I procured a ladder from my yard and climbed up to inspect the situation. There she was, the universe running down around her, warmly arranged among her guy ropes attached to the lamp supports—a great black and yellow embodiment of the life force, not giving up to either frost or stepladders. She ignored me and went on tightening and improving her web.

I stood over her on the ladder, a faint snow touching my cheeks, and surveyed her universe. There were a couple of iridescent green beetle cases turning slowly on a loose strand of web, a fragment of luminescent eye from a moth's wing and a large indeterminable object, perhaps a cicada, that had struggled and been wrapped in silk. There were also little bits and slivers, little red and blue flashes from the scales of anonymous wings that had crashed there.

Some days, I thought, they will be dull and gray and the shine will be out of them; then the dew will polish them again and drops hang on the silk until everything is gleaming and turning in the light. It is like a mind, really, where everything changes but remains, and in the end you have these eaten-out bits of experience like beetle wings.

I stood over her a moment longer, comprehending somewhat reluctantly that her adventure against the great blind forces of winter, her seizure of this warming globe of light, would come to nothing and was hopeless. Nevertheless it brought the birds back into my mind, and that faraway song which had traveled with growing strength around a forest clearing years ago—a kind of heroism, a world where even a spider refuses to lie down and die if a rope can still be spun on to a star. Maybe man himself will fight like this in the end, I thought, slowly realizing that the web and its threatening yellow occupant had been added to some luminous store of experience, shining for a moment in the fogbound reaches of my brain.

The mind, it came to me as I slowly descended the ladder, is a very remarkable thing; it has gotten itself a kind of courage by looking at a spider in a street lamp. Here was something that ought to be passed on to those who will fight our final freezing battle with the void. I thought of setting it down carefully as a message to the future: *In the days of the frost seek a minor sun.*

But as I hesitated, it became plain that something was wrong. The marvel was escaping—a sense of bigness beyond man's power to grasp, the essence of life in its great dealings with the universe. It was better, I decided, for the emissaries returning from the wilderness, even if they were merely descending from a stepladder, to record their marvel, not to define its meaning. In that way it would go echoing on through the minds of men, each grasping at that beyond out of which the miracles emerge, and which, once defined, ceases to satisfy the human need for symbols.

In the end I merely made a mental note: One specimen of Epeira observed building a web in a street light. Late autumn and cold for spiders. Cold for men, too. I shivered and left the lamp glowing there in my mind. The last I saw of Epeira she was hauling steadily on a cable. I stepped carefully over her shadow as I walked away.

THE BIRD
AND THE MACHINE

🌳

I suppose their little bones have years ago been lost among the stones and winds of those high glacial pastures. I suppose their feathers blew eventually into the piles of tumbleweed beneath the straggling cattle fences and rotted there in the mountain snows, along with dead steers and all the other things that drift to an end in the corners of the wire. I do not quite know why I should be thinking of birds over the *New York Times* at breakfast, particularly the birds of my youth half a continent away. It is a funny thing what the brain will do with memories and how it will treasure them and finally bring them into odd juxtapositions with other things, as though it wanted to make a design, or get some meaning out of them, whether you want it or not, or even see it.

It used to seem marvelous to me, but I read now that there are machines that can do these things in a small way, machines that can crawl about like animals, and that it may not be long now until they do more things —maybe even make themselves—I saw that piece in the *Times* just now. And then they will, maybe—well, who knows—but you read about it more and more with no one making any protest, and already they can add better than we and reach up and hear things through the dark and finger the guns over the night sky.

This is the new world that I read about at breakfast. This is the world that confronts me in my biological books and journals, until there are times when I sit quietly in my chair and try to hear the little purr of the cogs in my head and the tubes flaring and dying as the messages go through them and the circuits snap shut or open. This is the great age, make no mistake about it; the robot has been born somewhat appropriately along with the atom bomb, and the brain they say now is just another type of more complicated feedback system. The engineers have its basic principles worked out; it's mechanical, you know; nothing to get superstitious about; and man can always improve on nature once he gets the idea. Well, he's got it all right and that's why, I guess, that I sit here in my chair, with the article crunched in my hand, remembering those two birds and that blue mountain sunlight. There is another magazine article on my desk that reads "Machines Are Getting Smarter Every Day." I don't deny it, but I'll still stick with the birds. It's life I believe in, not machines.

Maybe you don't believe there is any difference. A skeleton is all joints and pulleys, I'll admit. And when man was in his simpler stages of machine building in the eighteenth century, he quickly saw the resemblances. "What," wrote Hobbes, "is the heart but a spring, and the nerves but so many strings, and the joints but so many wheels, giving motion to the whole body?" Tinkering about in their shops it was inevitable in the end that men would see the world as a huge machine "subdivided into an infinite number of lesser machines."

The idea took on with a vengeance. Little automatons toured the country—dolls controlled by clockwork. Clocks described as little worlds were taken on tours by their designers. They were made up of moving figures, shifting scenes and other remarkable devices. The life of the cell was unknown. Man, whether he was conceived as possessing a soul or not, moved and jerked about like these tiny puppets. A human being thought of himself in terms of his own tools and implements. He had been fashioned like the puppets he produced and was only a more clever model made by a greater designer.

Then in the nineteenth century, the cell was discovered, and the single machine in its turn was found to be the product of millions of infinitesimal machines —the cells. Now, finally, the cell itself dissolves away into an abstract chemical machine—and that into some intangible, inexpressible flow of energy. The secret seems to lurk all about, the wheels get smaller and smaller, and they turn more rapidly, but when you try to seize it the life is gone—and so, by popular definition, some would say that life was never there in the first place. The wheels and the cogs are the secret and we can make them better in time—machines that will run faster and more accurately than real mice to real cheese.

I have no doubt it can be done, though a mouse harvesting seeds on an autumn thistle is to me a fine sight and more complicated, I think, in his multiform activity, than a machine "mouse" running a maze. Also, I like to think of the possible shape of the future brooding in mice, just as it brooded once in a rather ordinary mousy insectivore who became a man. It leaves a nice fine indeterminate sense of wonder that even an electronic brain hasn't got, because you know perfectly well that if the electronic brain changes, it will be because of something man has done to it. But what man will do to himself he doesn't really know. A certain scale of time and a ghostly intangible thing called change are ticking in him. Powers and potentialities like the oak in the seed, or a red and awful ruin. Either way, it's impressive; and the mouse has it, too. Or those birds, I'll never forget those birds—yet before I measured their significance, I learned the lesson of time first of all. I was young then and left alone in a great desert —part of an expedition that had scattered its men over several hundred miles in order to carry on research more effectively. I learned there that time is a series of planes existing superficially in the same universe. The tempo is a human illusion, a subjective clock ticking in our own kind of protoplasm.

As the long months passed, I began to live on the slower planes and to observe more readily what passed for life there. I sauntered, I passed more and more slowly up and down the canyons in the dry baking heat of midsummer. I slumbered for long hours in the shade of huge brown boulders that had gathered in tilted companies out on the flats. I had forgotten the world of men and the world had forgotten me. Now and then I found a skull in the canyons, and these justified my remaining there. I took a serene cold interest in these discoveries. I had come, like many a naturalist before me, to view life with a wary and subdued attention. I had grown to take pleasure in the divested bone.

I sat once on a high ridge that fell away before me into a waste of sand dunes. I sat through hours of a long afternoon. Finally, as I glanced beside my boot an indistinct configuration caught my eye. It was a coiled rattlesnake, a big one. How long he had sat with me I do not know. I had not frightened him. We were both locked in the sleep-walking tempo of the earlier world, baking in the same high air and sunshine. Perhaps he had been there when I came. He slept on as I left, his coils, so ill discerned by me, dissolving once more among the stones and gravel from which I had barely made him out.

Another time I got on a higher ridge, among some tough little wind-warped pines half covered over with sand in a basin-like depression that caught everything carried by the air up to those heights. There were a few thin bones of birds, some cracked shells of indeterminable age, and the knotty fingers of pine roots bulged out of shape from their long and agonizing grasp upon the crevices of the rock. I lay under the pines in the sparse shade and went to sleep once more.

It grew cold finally, for autumn was in the air by then, and the few things that lived thereabouts were sinking down into an even chillier scale of time. In the moments between sleeping and waking I saw the roots about me and slowly, slowly, a foot in what seemed many centuries, I moved my sleep-stiffened hands over the scaling bark and lifted my numbed face after the vanishing sun. I was a great awkward thing of knots and aching limbs, trapped up there in some long, patient endurance that involved the necessity of putting living fingers into rock and by slow, aching expansion bursting those rocks asunder. I suppose, so thin and slow was the time of my pulse by then, that I might have stayed on to drift still deeper into the lower cadences of the frost, or the crystalline life that glistens pebbles, or shines in a snowflake, or dreams in the meteoric iron between the worlds.

It was a dim descent, but time was present in it. Somewhere far down in that scale the notion struck me that one might come the other way. Not many months thereafter I joined some colleagues heading higher into a remote windy tableland where huge bones were reputed to protrude like boulders from the turf. I had drowsed with reptiles and moved with the century-long pulse of trees; now, lethargically, I was climbing back up some invisible ladder of quickening hours. There had been talk of birds in connection with my duties. Birds are intense, fast-living creatures—reptiles, I suppose one might say, that have escaped out of the heavy sleep of time, transformed fairy creatures dancing over sunlit meadows. It is a youthful fancy, no doubt, but because of something that happened up there among the escarpments of that range, it remains with me a life-long impression. I can never bear to see a bird imprisoned.

We came into that valley through the trailing mists of a spring night. It was a place that looked as though it might never have known the foot of man, but our scouts had been ahead of us and we knew all about the abandoned cabin of stone that lay far up on one hillside. It had been built in the land rush of the last century and then lost to the cattlemen again as the marginal soils failed to take to the plow.

There were spots like this all over that country. Lost graves marked by unlettered stones and old corroding rim-fire cartridge cases lying where somebody had made a stand among the boulders that rimmed the valley. They are all that remain of the range wars; the men are under the stones now. I could see our cavalcade winding in and out through the mist below us: torches, the reflection of the truck lights on our collecting tins, and the far-off bumping of a loose dinosaur thigh bone in the bottom of a trailer. I stood on a rock a moment looking down and thinking what it cost in money and equipment to capture the past.

We had, in addition, instructions to lay hands on the present. The word had come through to get them alive —birds, reptiles, anything. A zoo somewhere abroad needed restocking. It was one of those reciprocal matters in which science involves itself. Maybe our museum needed a stray ostrich egg and this was the payoff. Anyhow, my job was to help capture some birds and that was why I was there before the trucks.

The cabin had not been occupied for years. We intended to clean it out and live in it, but there were holes in the roof and the birds had come in and were roosting in the rafters. You could depend on it in a place like this where everything blew away, and even a bird needed some place out of the weather and away from coyotes. A cabin going back to nature in a wild place draws them till they come in, listening at the eaves, I imagine, pecking softly among the shingles till they find a hole and then suddenly the place is theirs and man is forgotten.

Sometimes of late years I find myself thinking the most beautiful sight in the world might be the birds taking over New York after the last man has run away to the hills. I will never live to see it, of course, but I know just how it will sound because I've lived up high and I know the sort of watch birds keep on us. I've listened to sparrows tapping tentatively on the outside of air conditioners when they thought no one was listening, and I know how other birds test the vibrations that come up to them through the television aerials.

"Is he gone?" they ask, and the vibrations come up from below, "Not yet, not yet."

Well, to come back, I got the door open softly and I had the spotlight all ready to turn on and blind whatever birds there were so they couldn't see to get out through the roof. I had a short piece of ladder to put against the far wall where there was a shelf on which

I expected to make the biggest haul. I had all the information I needed just like any skilled assassin. I pushed the door open, the hinges squeaking only a little. A bird or two stirred—I could hear them—but nothing flew and there was a faint starlight through the holes in the roof.

I padded across the floor, got the ladder up and the light ready, and slithered up the ladder till my head and arms were over the shelf. Everything was dark as pitch except for the starlight at the little place back of the shelf near the eaves. With the light to blind them, they'd never make it. I had them. I reached my arm carefully over in order to be ready to seize whatever was there and I put the flash on the edge of the shelf where it would stand by itself when I turned it on. That way I'd be able to use both hands.

Everything worked perfectly except for one detail —I didn't know what kind of birds were there. I never thought about it at all, and it wouldn't have mattered if I had. My orders were to get something interesting. I snapped on the flash and sure enough there was a great beating and feathers flying, but instead of my having them, they, or rather he, had me. He had my hand, that is, and for a small hawk not much bigger than my fist he was doing all right. I heard him give one short metallic cry when the light went on and my hand descended on the bird beside him; after that he was busy with his claws and his beak was sunk in my thumb. In the struggle I knocked the lamp over on the shelf, and his mate got her sight back and whisked neatly through the hole in the roof and off among the stars outside. It all happened in fifteen seconds and you might think I would have fallen down the ladder, but no, I had a professional assassin's reputation to keep up, and the bird, of course, made the mistake of thinking the hand was the enemy and not the eyes behind it. He chewed my thumb up pretty effectively and lacerated my hand with his claws, but in the end I got him, having two hands to work with.

He was a sparrow hawk and a fine young male in the prime of life. I was sorry not to catch the pair of them, but as I dripped blood and folded his wings carefully, holding him by the back so that he couldn't strike again, I had to admit the two of them might have been more than I could have handled under the circumstances. The little fellow had saved his mate by diverting me, and that was that. He was born to it, and made no outcry now, resting in my hand hopelessly, but peering toward me in the shadows behind the lamp with a fierce, almost indifferent glance. He neither gave nor expected mercy and something out of the high air passed from him to me, stirring a faint embarrassment.

I quit looking into that eye and managed to get my huge carcass with its fist full of prey back down the ladder. I put the bird in a box too small to allow him to injure himself by struggle and walked out to welcome the arriving trucks. It had been a long day, and camp still to make in the darkness. In the morning that bird would be just another episode. He would go back with the bones in the truck to a small cage in a city where he would spend the rest of his life. And a good thing, too. I sucked my aching thumb and spat out some blood. An assassin has to get used to these things. I had a professional reputation to keep up.

In the morning, with the change that comes on suddenly in that high country, the mist that had hovered below us in the valley was gone. The sky was a deep blue, and one could see for miles over the high outcroppings of stone. I was up early and brought the box in which the little hawk was imprisoned out onto the grass where I was building a cage. A wind as cool as a mountain spring ran over the grass and stirred my hair. It was a fine day to be alive. I looked up and all around and at the hole in the cabin roof out of which the other little hawk had fled. There was no sign of her anywhere that I could see.

"Probably in the next county by now," I thought cynically, but before beginning work I decided I'd have a look at my last night's capture.

Secretively, I looked again all around the camp and up and down and opened the box. I got him right out in my hand with his wings folded properly and I was careful not to startle him. He lay limp in my grasp and

I could feel his heart pound under the feathers but he only looked beyond me and up.

I saw him look that last look away beyond me into a sky so full of light that I could not follow his gaze. The little breeze flowed over me again, and nearby a mountain aspen shook all its tiny leaves. I suppose I must have had an idea then of what I was going to do, but I never let it come up into consciousness. I just reached over and laid the hawk on the grass.

He lay there a long minute without hope, unmoving, his eyes still fixed on that blue vault above him. It must have been that he was already so far away in heart that he never felt the release from my hand. He never even stood. He just lay with his breast against the grass.

In the next second after that long minute he was gone. Like a flicker of light, he had vanished with my eyes full on him, but without actually seeing even a premonitory wing beat. He was gone straight into that towering emptiness of light and crystal that my eyes could scarcely bear to penetrate. For another long moment there was silence. I could not see him. The light was too intense. Then from far up somewhere a cry came ringing down.

I was young then and had seen little of the world, but when I heard that cry my heart turned over. It was not the cry of the hawk I had captured; for, by shifting my position against the sun, I was now seeing further up. Straight out of the sun's eye, where she must have been soaring restlessly above us for untold hours, hurtled his mate. And from far up, ringing from peak to peak of the summits over us, came a cry of such unutterable and ecstatic joy that it sounds down across the years and tingles among the cups on my quiet breakfast table.

I saw them both now. He was rising fast to meet her. They met in a great soaring gyre that turned to a whirling circle and a dance of wings. Once more, just once, their two voices, joined in a harsh wild medley of question and response, struck and echoed against the pinnacles of the valley. Then they were gone forever somewhere into those upper regions beyond the eyes of men.

I am older now, and sleep less, and have seen most of what there is to see and am not very much impressed any more, I suppose, by anything. "What Next in the Attributes of Machines?" my morning headline runs. "It Might Be the Power to Reproduce Themselves."

I lay the paper down and across my mind a phrase floats insinuatingly: "It does not seem that there is anything in the construction, constituents, or behavior of the human being which it is essentially impossible for science to duplicate and synthesize. On the other hand . . ."

All over the city the cogs in the hard, bright mechanisms have begun to turn. Figures move through computers, names are spelled out, a thoughtful machine selects the fingerprints of a wanted criminal from an array of thousands. In the laboratory an electronic mouse runs swiftly through a maze toward the cheese it can neither taste nor enjoy. On the second run it does better than a living mouse.

"On the other hand . . ." Ah, my mind takes up, on the other hand the machine does not bleed, ache, hang for hours in the empty sky in a torment of hope to learn the fate of another machine, nor does it cry out with joy nor dance in the air with the fierce passion of a bird. Far off, over a distance greater than space, that remote cry from the heart of heaven makes a faint buzzing among my breakfast dishes and passes on and away.

THE SECRET OF LIFE

✸

I am middle-aged now, but in the autumn I always seek for it again hopefully. On some day when the leaves are red, or fallen, and just after the birds are gone, I put on my hat and an old jacket, and over the protests of my wife that I will catch cold, I start my search. I go carefully down the apartment steps and climb, instead of jump, over the wall. A bit further I reach an unkempt field full of brown stalks and emptied seed pods.

By the time I get to the wood I am carrying all manner of seeds hooked in my coat or piercing my socks or sticking by ingenious devices to my shoestrings. I let them ride. After all, who am I to contend against such ingenuity? It is obvious that nature, or some part of it in the shape of these seeds, has intentions beyond this field and has made plans to travel with me.

We, the seeds and I, climb another wall together and sit down to rest, while I consider the best way to search for the secret of life. The seeds remain very quiet and some slip off into the crevices of the rock. A woolly-bear caterpillar hurries across a ledge, going late to some tremendous transformation, but about this he knows as little as I.

It is not an auspicious beginning. The things alive do not know the secret, and there may be those who would doubt the wisdom of coming out among discarded husks in the dead year to pursue such questions. They might say the proper time is spring, when one can consult the water rats or listen to little chirps under the stones. Of late years, however, I have come to suspect that the mystery may just as well be solved in a carved and intricate seed case out of which the life has flown, as in the seed itself.

In autumn one is not confused by activity and green leaves. The underlying apparatus, the hooks, needles, stalks, wires, suction cups, thin pipes, and iridescent bladders are all exposed in a gigantic dissection. These are the essentials. Do not be deceived simply because the life has flown out of them. It will return, but in the meantime there is an unparalleled opportunity to examine in sharp and beautiful angularity the shape of life without its disturbing muddle of juices and leaves. As I grow older and conserve my efforts, I shall give this season my final and undivided attention. I shall be found puzzling over the saw teeth on the desiccated leg of a dead grasshopper or standing bemused in a brown sea of rusty stems. Somewhere in this discarded machinery may lie the key to the secret. I shall not let it escape through lack of diligence or through fear of the smiles of people in high windows. I am sure now that life is not what it is purported to be and that nature, in the canny words of a Scotch theologue, "is not as natural as it looks." I have learned this in a small suburban field, after a good many years spent in much wilder places upon far less fantastic quests.

The notion that mice can be generated spontaneously from bundles of old clothes is so delightfully whimsical that it is easy to see why men were loath to abandon it. One could accept such accidents in a topsy-turvy universe without trying to decide what transformation of buckles into bones and shoe buttons into eyes had taken place. One could take life as a kind of fantastic magic and not blink too obviously when it appeared, beady-eyed and bustling, under the laundry in the back room.

It was only with the rise of modern biology and the discovery that the trail of life led backward toward infinitesimal beginnings in primordial sloughs, that men began the serious dissection and analysis of the cell. Darwin, in one of his less guarded moments, had spoken hopefully of the possibility that life had emerged from inorganic matter in some "warm little pond." From that day to this biologists have poured, analyzed, minced, and shredded recalcitrant protoplasm in a fruitless attempt to create life from nonliving matter. It seemed inevitable, if we could trace life down through simpler stages, that we must finally arrive at the point where, under the proper chemical conditions, the mysterious borderline that bounds the inanimate must be crossed. It seemed clear that life was a material manifestation. Somewhere, somehow, sometime, in the mysterious chemistry of carbon, the long march toward the talking animal had begun.

A hundred years ago men spoke optimistically about solving the secret, or at the very least they thought the next generation would be in a position to do so. Periodically there were claims that the emergence of life from matter had been observed, but in every case the observer proved to be self-deluded. It became obvious that the secret of life was not to be had by a little casual experimentation, and that life in today's terms appeared to arise only through the medium of preëxisting life. Yet, if science was not to be embarrassed by some kind of mind-matter dualism and a complete and irrational break between life and the world of inorganic matter, the emergence of life had, in some way, to be accounted for.

Nevertheless, as the years passed, the secret remained locked in its living jelly, in spite of larger microscopes and more formidable means of dissection. As a matter of fact the mystery was heightened because all this intensified effort revealed that even the supposedly simple amoeba was a complex, self-operating chemical factory. The notion that he was a simple blob, the discovery of whose chemical composition would enable us instantly to set the life process in operation, turned out to be, at best, a monstrous caricature of the truth.

With the failure of these many efforts science was left in the somewhat embarrassing position of having to postulate theories of living origins which it could not demonstrate. After having chided the theologian for his reliance on myth and miracle, science found itself in the unenviable position of having to create a mythology of its own: namely, the assumption that what, after long effort, could not be proved to take place today had, in truth, taken place in the primeval past.

My use of the term *mythology* is perhaps a little harsh. One does occasionally observe, however, a tendency for the beginning zoological textbook to take the unwary reader by a hop, skip, and jump from the little steaming pond or the beneficent chemical crucible of the sea, into the lower world of life with such sureness and rapidity that it is easy to assume that there is no mystery about this matter at all, or, if there is, that it is a very little one.

This attitude has indeed been sharply criticized by the distinguished British biologist Woodger, who remarked some years ago: "Unstable organic compounds and chlorophyll corpuscles do not persist or come into existence in nature on their own account at the present day, and consequently it is necessary to postulate that conditions were once such that this did happen although and in spite of the fact that our knowledge of nature does not give us any warrant for making such a supposition . . . It is simple dogmatism—asserting that what you want to believe did in fact happen."

Yet, unless we are to turn to supernatural explanations or reinvoke a dualism which is scientifically dubious, we are forced inevitably toward only two possible explanations of life upon earth. One of these, although not entirely disproved, is most certainly out of fashion and surrounded with greater obstacles to its acceptance than at the time it was formulated. I refer, of course, to the suggestion of Lord Kelvin and Svante Arrhenius that life did not arise on this planet, but was wafted here through the depths of space. Microscopic spores, it was contended, have great resistance to extremes of cold and might have come into our atmosphere with meteoric dust, or have been driven across the earth's orbit by light pressure. In this view, once the seed was "planted" in soil congenial to its development, it then proceeded to elaborate, evolve, and adjust until the higher organisms had emerged.

This theory had a certain attraction as a way out of an embarrassing dilemma, but it suffers from the defect of explaining nothing, even if it should prove true. It does not elucidate the nature of life. It simply removes the inconvenient problem of origins to far-off spaces or worlds into which we will never penetrate. Since life makes use of the chemical compounds of this earth, it would seem better to proceed, until incontrovertible evidence to the contrary is obtained, on the assumption that life has actually arisen upon this planet. The now widely accepted view that the entire universe in its present state is limited in time, and the apparently lethal nature of unscreened solar radiation are both obstacles which greatly lessen the likelihood that life has come to us across the infinite wastes of space. Once more, therefore, we are forced to examine our remaining notion that life is not coterminous with matter, but has arisen from it.

If the single-celled protozoans that riot in roadside pools are not the simplest forms of life, if, as we know today, these creatures are already highly adapted and really complex, though minute beings, then where are we to turn in the search for something simple enough to suggest the greatest missing link of all—the link between living and dead matter? It is this problem that keeps me wandering fruitlessly in pastures and weed thickets even though I know this is an old-fashioned naturalist's approach, and that busy men in laboratories have little patience with my scufflings of autumn leaves, or attempts to question beetles in decaying bark. Besides, many of these men are now fascinated by the crystalline viruses and have turned that remarkable instrument, the electron microscope, upon strange molecular "beings" never previously seen by man. Some are satisfied with this glimpse below the cell and find the virus a halfway station on the road to life. Perhaps it is, but as I wander about in the thin mist that is beginning to filter among these decaying stems and ruined spider webs, a kind of disconsolate uncertainty has taken hold of me.

I have come to suspect that this long descent down the ladder of life, beautiful and instructive though it may be, will not lead us to the final secret. In fact I have ceased to believe in the final brew or the ultimate chemical. There is, I know, a kind of heresy, a shocking negation of our confidence in blue-steel microtomes and men in white in making such a statement. I would not be understood to speak ill of scientific effort, for in simple truth I would not be alive today except for the microscopes and the blue steel. It is only that somewhere among these seeds and beetle shells and abandoned grasshopper legs I find something that is not accounted for very clearly in the dissections to the ultimate virus or crystal or protein particle. Even if the secret is contained in these things, in other words, I do not think it will yield to the kind of analysis our science is capable of making.

Imagine, for a moment, that you have drunk from a magician's goblet. Reverse the irreversible stream of time. Go down the dark stairwell out of which the race has ascended. Find yourself at last on the bottommost steps of time, slipping, sliding, and wallowing by scale and fin down into the muck and ooze out of which you arose. Pass by grunts and voiceless hissings below the last tree ferns. Eyeless and earless, float in the primal waters, sense sunlight you cannot see and stretch absorbing tentacles toward vague tastes that float in water. Still, in your formless shiftings, the *you* remains: the sliding particles, the juices, the transformations are working in an exquisitely patterned rhythm which has no other purpose than your preservation—you, the entity, the ameboid being whose substance contains the unfathomable future. Even so does every man come upward from the waters of his birth.

Yet if at any moment the magician bending over you should cry, "Speak! Tell us of that road!" you could not respond. The sensations are yours but not—and this is one of the great mysteries—the power over the body. You cannot describe how the body you inhabit functions, or picture or control the flights and spinnings, the dance of the molecules that compose it, or why they chose to dance into that particular pattern which is you, or, again, why up the long stairway of the eons they dance from one shape to another. It is for this reason that I am no longer interested in final particles. Follow them as you will, pursue them until they become nameless protein crystals replicating on the verge of life. Use all the great powers of the mind and pass backward until you hang with the dire faces of the conquerors in the hydrogen cloud from which the sun was born. You will then have performed the ultimate dissection that our analytic age demands, but the cloud will still veil the secret and, if not the cloud, then the nothingness into which, it now appears, the cloud, in its turn, may be dissolved. The secret, if one may paraphrase a savage vocabulary, lies in the egg of night.

Only along the edges of this field after the frost there are little whispers of it. Once even on a memorable autumn afternoon I discovered a sunning blacksnake brooding among the leaves like the very simulacrum of old night. He slid unhurriedly away, carrying his version of the secret with him in such a glittering menace of scales that I was abashed and could only follow admiringly from a little distance. I observed him well, however, and am sure he carried his share of the common mystery into the stones of my neighbor's wall, and is sleeping endlessly on in the winter darkness with one great coil locked around that glistening head. He is guarding a strange, reptilian darkness which is not night or nothingness, but has, instead, its momentary vision of mouse bones or a bird's egg, in the soft rising and ebbing of the tides of life. The snake has diverted me, however. It was the dissection of a field that was to occupy us—a dissection in search of secrets—a dissection such as a probing and inquisitive age demands.

Every so often one encounters articles in leading magazines with titles such as "The Spark of Life," "The Secret of Life," "New Hormone Key to Life," or other similar optimistic proclamations. Only yesterday, for example, I discovered in the *New York Times* a headline announcing: "Scientist Predicts Creation of Life in Laboratory." The Moscow-date-lined dispatch announced that Academician Olga Lepeshinskaya had predicted that "in the not too distant future, Soviet scientists would create life." "The time is not far off," warns the formidable Madame Olga, "when we shall be able to obtain the vital substance artificially." She said it with such vigor that I had about the same reaction as I do to announcements about atomic bombs. In fact I half started up to latch the door before an invading tide of Russian protoplasm flowed in upon me.

What finally enabled me to regain my shaken confidence was the recollection that these pronouncements have been going on for well over a century. Just now the Russian scientists show a particular tendency to issue such blasts—committed politically, as they are, to an uncompromising materialism and the boastfulness of very young science. Furthermore, Madame Lepeshinskaya's remarks as reported in the press had a curiously old-fashioned flavor about them. The protoplasm she referred to sounded amazingly like the outmoded *Urschleim* or *Autoplasson* of Haeckel—simplified mucoid slimes no longer taken very seriously. American versions—and one must remember they are often journalistic interpretations of scientists' studies rather than direct quotations from the scientists themselves— are more apt to fall into another pattern. Someone has found a new chemical, vitamin, or similar necessary ingredient without which life will not flourish. By the time this reaches the more sensational press, it may have become the "secret of life." The only thing the inexperienced reader may not comprehend is the fact that no one of these items, even the most recently discovered, is *the* secret. Instead, the substance is probably a part, a very small part, of a larger enigma which is well-nigh as inscrutable as it ever was. If anything, the growing list of catalysts, hormones, plasma genes, and other hobgoblins involved in the work of life only serves to underline the enormous complexity of the secret. "To grasp in detail," says the German biologist Von Bertalanffy, "the physico-chemical organization of the simplest cell is far beyond our capacity."

It is not, you understand, disrespect for the laudable and persistent patience of these dedicated scientists happily lost in their maze of pipettes, smells, and gas flames, that has led me into this runaway excursion to the wood. It is rather the loneliness of a man who knows he will not live to see the mystery solved, and who, furthermore, has come to believe that it will not be solved when the first humanly synthesized particle begins—if it ever does—to multiply itself in some unknown solution.

It is really a matter, I suppose, of the kind of questions one asks oneself. Some day we may be able to say with assurance, "We came from such and such a protein particle, possessing the powers of organizing in a manner leading under certain circumstances to that complex entity known as the cell, and from the cell by various steps onward, to multiple cell formation." I mean we may be able to say all this with great surety and elaboration of detail, but it is not the answer to the grasshopper's leg, brown and black and saw-toothed here in my hand, nor the answer to the seeds still clinging tenaciously to my coat, nor to this field, nor to the subtle essences of memory, delight, and wistfulness moving among the thin wires of my brain.

I suppose that in the forty-five years of my existence every atom, every molecule that composes me has changed its position or danced away and beyond to become part of other things. New molecules have come from the grass and the bodies of animals to be part of me a little while, yet in this spinning, light and airy as a midge swarm in a shaft of sunlight, my memories hold, and a loved face of twenty years ago is before me still. Nor is that face, nor all my years, caught cellularly as in some cold precise photographic pattern, some gross, mechanical reproduction of the past. My memory holds the past and yet paradoxically knows, at the same time, that the past is gone and will never come again. It cherishes dead faces and silenced voices, yes, and lost evenings of childhood. In some odd nonspatial way it contains houses and rooms that have been torn timber from timber and brick from brick. These have a greater permanence in that midge dance which contains them than ever they had in the world of reality. It is for this reason that Academician Olga Lepeshinskaya has not answered the kind of questions one may ask in an open field.

If the day comes when the slime of the laboratory for the first time crawls under man's direction, we shall have great need of humbleness. It will be difficult for us to believe, in our pride of achievement, that the secret of life has slipped through our fingers and eludes us still. We will list all the chemicals and the reactions. The men who have become gods will pose austerely before the popping flashbulbs of news photographers, and there will be few to consider—so deep is the mind-set of an age—whether the desire to link life to matter may not have blinded us to the more remarkable characteristics of both.

As for me, if I am still around on that day, I intend to put on my old hat and climb over the wall as usual. I shall see strange mechanisms lying as they lie here now, in the autumn rain, strange pipes that transported the substance of life, the intricate seedcase out of which the life has flown. I shall observe no thing green, no delicate transpirations of leaves, nor subtle comings and goings of vapor. The little sunlit factories of the chloroplasts will have dissolved away into common earth.

Beautiful, angular, and bare the machinery of life will lie exposed, as it now is, to my view. There will be the thin, blue skeleton of a hare tumbled in a little heap, and crouching over it I will marvel, as I marvel now, at the wonderful correlation of parts, the perfect adaptation to purpose, the individually vanished and yet persisting pattern which is now hopping on some other hill. I will wonder, as always, in what manner "particles" pursue such devious plans and symmetries. I will ask once more in what way it is managed, that the simple dust takes on a history and begins to weave these unique and never recurring apparitions in the stream of time. I shall wonder what strange forces at the heart of matter regulate the tiny beating of a rabbit's heart or the dim dream that builds a milkweed pod.

It is said by men who know about these things that the smallest living cell probably contains over a quarter of a million protein molecules engaged in the multitudinous coördinated activities which make up the phenomenon of life. At the instant of death, whether of man or microbe, that ordered, incredible spinning passes away in an almost furious haste of those same particles to get themselves back into the chaotic, unplanned earth.

I do not think, if someone finally twists the key successfully in the tiniest and most humble house of life, that many of these questions will be answered, or that the dark forces which create lights in the deep sea and living batteries in the waters of tropical swamps, or the dread cycles of parasites, or the most noble workings of the human brain, will be much if at all revealed. Rather, I would say that if "dead" matter has reared up this curious landscape of fiddling crickets, song sparrows, and wondering men, it must be plain even to the most devoted materialist that the matter of which he speaks contains amazing, if not dreadful powers, and may not impossibly be, as Hardy has suggested, "but one mask of many worn by the Great Face behind."

A native of Lincoln, Nebraska, LOREN EISELEY was born into a family which homesteaded in that region when it was still a territory. His first contact with nature lay in the salt flats and ponds around Lincoln, and in the mammoth bones hoarded in the old red brick museum on the campus of the University of Nebraska. Receiving his A.B. degree there, he completed graduate work in anthropology at the University of Pennsylvania. Returning to the Midwest for his first academic job, he taught at the University of Kansas. He later became head of the Department of Sociology and Anthropology at Oberlin College in Ohio, then returned to the University of Pennsylvania in 1947 to head the Department of Anthropology. He also is Curator of Early Man in the University Museum.

Dr. Eiseley has lectured at a number of universities, including Harvard, Columbia, and the University of California. He is past president of the American Institute of Human Paleontology, and a contributor to many leading scientific journals as well as periodicals such as *Harper's*, *American Scholar*, and *Gentry*.

For a number of years he was active in the search for early postglacial man in the western United States, and has worked extensively in the high plains, mountains, and deserts bordering the Rocky Mountains from Canada into Mexico.

VINTAGE FICTION, POETRY, AND PLAYS